Use of phosphate rocks for sustainable agriculture

FAO
FERTILIZER
AND PLANT
NUTRITION
BULLETIN
13

Technical editing by
F. Zapata
Joint FAO/IAEA Division of Nuclear Techniques in Food and Agriculture
Vienna, Austria

R.N. Roy
Land and Water Development Division
FAO, Rome, Italy

A joint publication of the
FAO Land and Water Development Division
and
the International Atomic Energy Agency

FOOD AND AGRICULTURE ORGANIZATION OF THE UNITED NATIONS
Rome, 2004

The designations employed and the presentation of material in this information
product do not imply the expression of any opinion whatsoever on the part
of the Food and Agriculture Organization of the United Nations or the International
Atomic Energy Agency concerning the legal or development status of any country,
territory, city or area or of its authorities, or concerning the delimitation of its frontiers
or boundaries.

ISBN 92-5-105030-9

All rights reserved. Reproduction and dissemination of material in this information
product for educational or other non-commercial purposes are authorized without
any prior written permission from the copyright holders provided the source is fully
acknowledged. Reproduction of material in this information product for resale or other
commercial purposes is prohibited without written permission of the copyright holders.
Applications for such permission should be addressed to:
Chief
Publishing Management Service
Information Division
FAO
Viale delle Terme di Caracalla, 00100 Rome, Italy
or by e-mail to:
copyright@fao.org

© FAO 2004

Contents

List of tables	vi
List of figures	viii
Acknowlegements	x
Preface	xi
Contributing authors	xiii
List of abbreviations and acronyms	xiv
Executive summary	xvii

1. Introduction — 1
- Phosphorus in the soil-plant system — 1
- The need for sustainable development — 2
- Addressing the P constraint in tropical acid soils — 3
- Use of phosphate rocks in industry and agriculture — 3
  - Phosphate rocks as raw materials for P-fertilizer manufacturing — 3
  - Phosphate rock for direct application in agriculture — 4
  - Historical account of the use of phosphate rocks in agriculture — 6
- The bulletin — 8

2. World phosphate deposits — 11
- World phosphate rock production — 11
- World phosphate rock reserves and resources — 12
- Future trends in world phosphate rock production — 15

3. Characterization of phosphate rocks — 17
- Phosphate rock mineralogy — 17
  - Sedimentary apatites — 17
  - Igneous apatites — 19
  - Other minerals in PRs — 19
- Phosphate rock solubility tests — 21
  - Apatite solubility — 21
- Classification of phosphate rocks based on solubility — 23
- Characterization methods — 24

4. Evaluation of phosphate rocks for direct application — 27
- Phosphate-rock solubility tests — 28
  - Solubility measurements using conventional chemical techniques — 28
  - Kinetics of long-term PR dissolution — 30
  - PR reactivity scales and crop yield response — 31
  - Measurement of exchangeable P from PR by radioisotopic techniques — 33
- Reactions between phosphate rocks and soil — 34
  - Soil incubation — 34
  - Liming effect of PRs — 35
- Greenhouse tests — 35
  - Comparison of PRs with standard fertilizers — 37

Effect of time	38
Relationship between PR solubility and crop P uptake	39
Field evaluation	39

5. Factors affecting the agronomic effectiveness of phosphate rocks, with a case-study analysis — 41

Factors affecting the agronomic effectiveness of phosphate rocks	41
Reactivity of PRs	41
Soil properties	42
Climate conditions	45
Crop species	45
Management practices	46
Case studies	47
Mali	47
Madagascar	49
India	50
Indonesia	52
New Zealand	54
Latin America	54

6. Soil testing for phosphate rock application — 59

Soil phosphorus reaction and soil testing	59
Conventional soil tests	60
Bray I test	60
Bray II and Mehlich I tests	62
Olsen test	63
Recently developed soil tests	64
Iron-oxide-impregnated paper test	64
Mixed anion and cation exchange resins	66
Isotopic ^{32}P exchange kinetic method	67

7. Decision-support systems for the use of phosphate rocks — 69

The need for a decision-support system for phosphate rocks	69
Conceptual basis for building a PR-DSS	69
Different types of DSS to predict PR performance	71
Mechanistic models	71
The combination of mechanistic and empirical models	71
Expert systems	72
The 'RPR Adviser' in Australia	72
Requirements for a global DSS for PR use	74
Creation of a database for PR reactivities	74
Creating a database of soils for PR use	74
Other requirements	75
Future developments	75

8. Secondary nutrients, micronutrients, liming effect and hazardous elements associated with phosphate rock use — 77

Secondary nutrients in phosphate rock	77
Micronutrients in phosphate rock	79
Liming effect associated with PR use	80
Hazardous elements in phosphate rock	82

9. WAYS OF IMPROVING THE AGRONOMIC EFFECTIVENESS OF PHOSPHATE ROCKS	85
Biological means	85
Phospho-composts	85
Inoculation of seedlings with endomycorrhizae	88
Use of ectomycorrhizae	90
Use of phosphate solubilizing micro-organisms	90
Use of plant genotypes	91
Chemical means	92
Partial acidulation of phosphate rock	92
Physical means	95
Compacted PR with water-soluble phosphate products	95
Dry mixtures of PR with water-soluble phosphate fertilizers	95
Phosphate rock elemental sulphur assemblages	95
10. ECONOMIC FACTORS IN THE ADOPTION AND UTILIZATION OF PHOSPHATE ROCKS	99
Possible adoption and use of phosphate rock by end users	100
Cost of production, transport and distribution	102
Economic considerations and policies to support PR adoption	104
Pricing policy	105
Organizational policy	105
Agricultural product markets	105
Risk protection	106
Research and extension policy	106
Case studies of PR exploitation and use	107
Marketing of indigenous phosphate rocks for direct application	109
Latin America	110
Asia	110
Sub-Saharan Africa	111
Economic criteria in PR adoption and use – a case study on Venezuela	111
Agronomic evaluation and domestic market potential	112
Economics of PAPR production	112
Pilot-plant evaluation test	114
PAPR quality	115
The Moron PAPR plant	115
Raw materials and products	115
11. LEGISLATION AND QUALITY CONTROL OF PHOSPHATE ROCKS FOR DIRECT APPLICATION	117
Current legislation on phosphate rock for direct application	117
Associated issues	118
Total P_2O_5 content of PR	118
Solubility of phosphate rock	119
Expressing PR solubility	119
Free carbonate effect	120
Particle-size effect	121
Guidelines for legislation on DAPR	122
12. EPILOGUE	125
BIBLIOGRAPHY	127

List of tables

1.	World phosphate rock production, 1999	12
2.	World consumption of direct application phosphate rock	12
3.	World phosphate rock reserves and reserve base	13
4.	World economic identified phosphate concentrate resources	14
5.	Solubility data on selected phosphate rock samples	21
6.	Proposed classification of PR for direct application by solubility and expected initial response	23
7.	Ranking system for some South American PRs by solubility and RAE	23
8.	Chemical analysis and solubility in conventional reagents of selected African PRs	28
9.	Citrate soluble P of North Carolina PR as affected by particle size	29
10.	Correlation coefficients between reactivity scales of seven PRs and crop yield	32
11.	Spatial variability of PR samples within the same deposit	32
12.	PR kinetics of isotopic exchange and solubility in water	33
13.	Effects of incubating PRs on the characteristics of an Oxisol from Njole, Gabon	34
14.	Results of a pot experiment with selected African PRs	36
15.	Estimated RAE coefficients based on L values and P uptake from the pot experiment	38
16.	Changes with time of the estimated RAE coefficients for PRs applied to a Madagascar Andosol	38
17.	P uptake from PRs in a Madagascar Oxisol as influenced by soil water content	39
18.	Correlation coefficients between P uptake and solubility tests	39
19.	Crop yield and estimated RAEs from field experiments in Burkina Faso	40
20.	Yield of millet, groundnut, sorghum, cotton and maize with Tilemsi PR and TSP, Mali, 1982–87	48
21.	Yield of rice and maize with Mussoorie PR, SSP and a 1:1 mixture of Mussoorie PR-SSP	51
22.	Net returns on P fertilizers to finger millet and maize, and black gram grown on the residual effect	52
23.	Relative agronomic effectiveness of PRs for crops on a Typic Hapludult, Pelaihari, Kalimantan	53
24.	Comparison between a rehabilitated pasture using the PAPR-urea-Stylosanthes capitata treatment and a traditional degraded pasture, Monagas State, Venezuela	56
25.	AEI of phosphate fertilizers in a clayey Oxisol of central Brazil, based on P uptake data over five years with annual crops, followed by three years with Andropogon pasture	56

26.	Relative agronomic effectiveness of various PRs with respect to $CaCO_3$ as a Ca source for maize	78
27.	Chemical analysis of potentially hazardous elements in sedimentary phosphate rocks	82
28.	Nutrient content of organic waste materials and farmyard compost (dry basis except for urine)	86
29.	Effect of P sources on yield and P uptake for red gram and clusterbean	88
30.	Matrix of costs and benefits associated with PR investments	108
31.	Distribution of benefits for selected fertilizer strategies	109
32.	Partially acidulated phosphate rocks based on various Venezuelan deposits	112
33.	Current production capacity for P_2O_5 at the Moron Complex	113
34.	Pilot-plant evaluation results	115
35.	Total P_2O_5, NAC-soluble P_2O_5, and CO_3:PO_4 ratio of apatite in various phosphate rocks	118
36.	Citrate solubility of mixtures of North Carolina PR and sand	120
37.	Citrate solubility of PRs containing various amounts of total P_2O_5	120
38.	Solubility of PRs as measured by various chemical extractions	121
39.	Solubility of ground and unground PRs	121

List of figures

1. Economic and potentially economic phosphate deposits of the world — 13
2. History of the discovery of world phosphate resources — 15
3. Excess fluorine francolites, correlation of moles of CO_3^{-2} with PO_4^{-3} per formula weight — 18
4. Excess fluorine francolites, variation in unit-cell a value with increasing CO_3^{-2} for PO_4^{-3} substitution — 18
5. Variation of CO_2 contents in excess fluorine francolites of various geological ages — 19
6. Relationship of the CaO/P_2O_5 weight percent ratio to NAC-soluble P_2O_5, excess fluorine francolites — 20
7. Unit-cell a dimension versus NAC solubility of hydroxyl-fluor-carbonate apatites — 20
8. NAC solubilities of sedimentary PRs from various countries — 22
9. Flowsheet of the PR characterization process at the IFDC — 24
10. Diagram showing the approximate modal analysis of phosphorite sample No. 1 — 26
11. Kinetics of continuous dissolution of PRs with formic acid — 31
12. Schematic representation of a crop response curve on adding PR to a severely P-deficient soil — 44
13. Effect of partial acidulation of Mussoorie PR on yields of rice (soil pH 7.9) and wheat (soil pH 6.0) — 51
14. Relationship between plant P uptake and soil available P from TSP and PR applied to acid soil — 60
15. Relationship between dry-matter yield and Bray I-extractable P in soils treated with PRs and TSP — 61
16. Relationship between dry-matter yield of maize with different P sources and Bray-I-P extracted from (a) acid sandy soil (pH 4.8) with low pH buffering capacity and (b) acid soil (pH 5.2) with high P-fixing capacity — 62
17. Relationship between dry-matter yield of maize and (a) Bray-II-extractable P, and (b) Mehlich-I-extractable P in soil treated with central Florida PR and TSP — 63
18. Relationship between pasture yield and Olsen P values in soils treated with Sechura PR and: (a) monocalcium phosphate, and (b) TSP — 64
19. Relationship between P uptake by maize and Bray I-extractable P in soils treated with TSP and North Carolina PR — 65
20. Relationship between P uptake by maize and Pi-P (0.01 M $CaCl_2$) in soils treated with TSP and North Carolina PR — 65
21. Pi-P with 0.01 M $CaCl_2$ or 0.02 M KCl in soils treated with North Carolina PR — 66

22.	Relationship between P uptake by maize and Pi-P (0.02 M KCl) in soils treated with TSP and North Carolina PR	66
23.	Pi-P with 0.01 M $CaCl_2$ or 0.02 M KCl in soils treated with TSP	67
24.	Process determining the feasibility of using PR for direct application	70
25.	Relationship between maximum P concentration in soil solution and mole ratio of CO_3:PO_4 in the apatite structure	78
26.	Relationship between Ca uptake by maize and citrate solubility of various PR sources	78
27.	Response of rainfed rice to Huila PR and TSP on an Oxisol	79
28.	Effect of Sechura PR and TSP on Mo concentration in clover	80
29.	Exchangeable Al and Ca in an Oxisol treated with PRs and TSP at 200 mg P/kg during incubation	81
30.	Effect of P sources on Panicum maximum dry-matter yield (sum of three cuts) on an Oxisol	81
31.	Relationship between the level of acidulation with (A) phosphoric acid and (B) sulphuric acid of North Carolina PR and the total and water-soluble P of the products	93
32.	Production synergies between the PAPR and phosphoric acid	114
33.	Production process flow chart of the PAPR plant	116

Acknowlegements

Specialists working on the development and use of phosphate rocks contributed chapters to this publication. Their contributions are greatly appreciated and highly acknowledged.

In addition to writing individual assigned chapters, most of the contributors also reviewed chapters written by other contributors. Thanks are due to all of them for their efforts in peer-reviewing the document.

Special thanks are due to R.N. Roy and F. Zapata for the conceptualization, initiation, technical guidance, inputs, reviewing and editing of this publication. Thanks are also extended to R.V. Misra and I. Verbeke for their inputs in finalizing this document.

Preface

This publication deals with the direct application of phosphate rock (PR) sources to agriculture. Phosphorus (P) is an essential plant nutrient and its deficiency restricts crop yields severely. Tropical and subtropical soils are predominantly acidic, and often extremely P deficient with high P-sorption (fixation) capacities. Therefore, substantial P inputs are required for optimum plant growth and adequate food and fibre production.

Manufactured water-soluble P fertilizers such as superphosphates are commonly recommended to correct P deficiencies, but most developing countries import these fertilizers, which are often in limited supply and represent a major outlay for resource-poor farmers. In addition, intensification of agricultural production in these countries necessitates the addition of P not only to increase crop production but also to improve soil P status in order to avoid further soil degradation. Hence, it is imperative to explore alternative P sources. Under certain soil and climate conditions, the direct application of PR, especially where available locally, has proved to be an agronomically and economically sound alternative to the more expensive superphosphates. PR deposits occur worldwide, but few are mined (for use mainly as raw materials to manufacture water-soluble P fertilizers).

The Joint FAO/IAEA Division of Nuclear Techniques in Food and Agriculture initiated a Coordinated Research Project called "The use of nuclear and related techniques for evaluating the agronomic effectiveness of phosphatic fertilizers, in particular rock phosphates". This was implemented by institutes of developing and industrialized countries from 1993 to 1998. The results obtained yielded new information on: chemistry of soil P; tests for available soil P; phosphate nutrition of crops; agronomic effectiveness of PR products; and P fertilizer recommendations with particular emphasis on PR use.

Within the framework of the integrated plant nutrition systems promoted by the Land and Water Development Division (AGL), FAO, and the national action plans for soil productivity improvement under the Soil Fertility Initiative for sub-Saharan countries, PRs are considered as potentially important locally available P sources. AGL has instituted several studies on the agro-economic assessment of PRs for direct application in selected countries. Results of practical utility and policy guidelines can be drawn from these studies.

Several organizations have conducted extensive research on the utilization of indigenous PR deposits in tropical soils in Latin America, Africa, Asia and elsewhere. In the past decade, considerable progress has been made in the utilization of PR sources for direct application in agricultural cropping systems worldwide. A wealth of information is now available but scattered in several publications.

Recognizing the need for the wider dissemination of the available information, the Joint FAO/IAEA Division of Nuclear Techniques in Food and Agriculture together with the AGL convened a Consultants' Meeting in Vienna in November 2001 in order to review advances in this field of research and development, and to elaborate a proposal for the production of this technical bulletin. Mr F. Zapata (IAEA, Vienna) and Mr R.N. Roy (FAO, Rome) implemented this task. Specialists in the sector were invited to contribute to the chapters of the publication.

This publication presents the results of studies on the utilization of PR products for direct application in agriculture under a wide range of agro-ecological conditions with a view to fostering sustainable agricultural intensification in developing countries of the tropics and subtropics. The subject matter is covered in 12 chapters, contributed by a team of scientists involved in PR research and working in a wide range of disciplines including geology, chemistry, soil science, agronomy and economics, etc. In addition, the publication includes a comprehensive bibliography.

Chapter 1 provides background information on: P as an essential plant nutrient; soil P status; the need for the application of P fertilizers; and the potential for using PRs in the context of food security, soil degradation and environmental protection. It describes generalities on the nature and variability of PR sources worldwide and provides an overview of past and current work and of the prospects for the utilization of PR products for direct application in agriculture. Chapter 2 deals with the type and distribution of PR geological deposits worldwide. It provides information on: inventory of reserves; PR production; and PR consumption for direct application. Chapter 3 describes the characterization of PR products in view of their wide variability for utilization as a raw material for the manufacture of P fertilizers and for direct application in agriculture. This includes information on: mineralogical composition of the phosphate-bearing minerals (main and accessory), in particular, the crystallographic features and empirical formulae of the apatite minerals; chemical composition (main chemical elements including micro-elements, heavy metals and radionuclides); solubility indices as indicators of reactivity; and physical properties (particle-size distribution, specific surface area, and hardness). Chapter 4 examines the approaches and methodologies for evaluating PR for direct application, such as solubility tests, soil incubation, greenhouse tests and field evaluation. For each approach, the chapter provides an overview of the objectives, methods and measurements with examples. It also provides a brief description of the advantages and limitations of the methods, and special considerations for interpreting results. Chapter 5 provides an up-to-date review of the main factors that affect the agronomic effectiveness of PR products. Selected case studies provide region-level examples of the influence of these factors and their interactions, and economic guidelines. Chapter 6 describes soil P testing methods for providing recommendations for PR application. It discusses principles and considerations for interpreting the results of 'available P'. Chapter 7 focuses on the need for and the development and testing of a decision-support system to predict the agronomic effectiveness of PR for direct application, together with the use of P modelling. Chapter 8 examines current knowledge, and gaps therein, on the effects of other elements present in PR products. It covers beneficial and hazardous elements, such as calcium and magnesium, micronutrients, heavy metals and radionuclides. Chapter 9 describes approaches and technologies for improving the agronomic effectiveness of indigenous PRs. Chapter 10 examines the economic criteria to be considered in the production, marketing and distribution of PRs. It discusses macroeconomic and microeconomic issues, and includes case studies on the exploitation of PR deposits. Chapter 11 provides an overview of current regulations on PR for direct application. Legislative guidelines are proposed for consideration. Chapter 12 is an epilogue containing the main conclusions and issues for further consideration.

The aim of this publication is to provide the most up-to-date information on the direct application of PR in agriculture. It endeavours to do this in a technically focused document that leads on to practical guidelines to assist middle-level decision-makers, the scientific community, higher-level extension workers, non-governmental organizations and other stakeholders involved in agricultural development. The ultimate goal is to maximize technology dissemination in a targeted way, particularly for policy-makers at all levels who may require information on the adoption of PRs as a capital investment to trigger sustainable agricultural intensification in the acidic soils of the tropics and subtropics.

Contributing authors

Chapter 1	Introduction, by F. Zapata (FAO/IAEA, Austria) and R.N. Roy (FAO, Italy)
Chapter 2	World phosphate deposits, by G.H. McClellan (University of Florida, the United States of America) and S.J. Van Kauwenbergh (IFDC, the United States of America)
Chapter 3	Characterization of phosphate rocks, by S.J. Van Kauwenbergh (IFDC, the United States of America) and G.H. McClellan (University of Florida, the United States of America)
Chapter 4	Evaluation of phosphate rocks for direct application, by B. Truong (France);
Chapter 5	Factors affecting the agronomic effectiveness of phosphate rocks, with a case-study analysis, by S.S.S. Rajan (University of Waikato, New Zealand), E. Casanova (Universidad Central de Venezuela, Venezuela) and B. Truong (France)
Chapter 6	Soil testing for phosphate rock application, by S.H. Chien (IFDC, the United States of America)
Chapter 7	Decision-support systems for the use of phosphate rocks, by P.W.G. Sale (La Trobe University, Australia) and L.K. Heng (FAO/IAEA, Austria)
Chapter 8	Secondary nutrients, micronutrients, liming effect, and hazardous elements associated with phosphate rock use, by S.H. Chien (IFDC, the United States of America)
Chapter 9	Ways of improving the agronomic effectiveness of phosphate rocks, by S.S.S. Rajan (University of Waikato, New Zealand)
Chapter 10	Economic factors in the adoption and utilization of phosphate rocks, by A. Kuyvenhoven (Wageningen University, the Netherlands), with a section on the marketing of indigenous phosphate rocks for direct application, by S.H. Chien (IFDC, the United States of America), and a section on economic criteria in adoption and use of phosphate rocks, including production, marketing and distribution – a case study in Venezuela, by E. Casanova (Universidad Central de Venezuela, Venezuela)
Chapter 11	Legislation and quality control of phosphate rocks for direct application, by S.H. Chien (IFDC, the United States of America)
Chapter 12	Epilogue, by R.N. Roy (FAO, Italy) and F. Zapata (FAO/IAEA, Austria)

List of abbreviations and acronyms

^{32}P	Phosphorus-32 (radioisotope)
AEI	Agronomic economic efficiency
AGL	Land and Water Development Division, Agriculture Department, FAO
Al	Aluminium
As	Arsenic
ASI	Absolute solubility index
BNF	Biological nitrogen fixation
BPL	Bone phosphate of lime
BRGM	Bureau of Geological and Mining Research (France)
C	Carbon
Ca	Calcium
CA	Citric acid
CCE	Calcium carbonate equivalent
Cd	Cadmium
CGIAR	Consultative Group on International Agricultural Research
CIRAD	Centre de Cooperation Internationale en Recherche Agronomique pour le Developpement
Cl	Chlorine
CMDT	Malian Company for the Development of Textiles
CO_2	Carbon dioxide
C_p	P ion concentration in soil solution
Cr	Chromium
CR	Equilibrium P concentration sustained by a PR
CRP	Coordinated research project
Cu	Copper
DAP	Di-ammonium phosphate
DAPR	Direct application phosphate rock
Dm	Mean diffusion
DSS	Decision-support system
DSSAT	Decision Support System for Agrotechnology Transfer
E value	Exchangeable P
E_1	Isotopically exchangeable P at 1 minute
EC	European Community
EDS	Energy dispersive X-ray analysis
F	Fluorine

FA	Formic acid
FAO/IAEA	Joint FAO/IAEA Division of Nuclear Techniques in Food and Agriculture
Fe	Iron
FSU	Former Soviet Union
H	Hydrogen
Hg	Mercury
IAEA	International Atomic Energy Agency
ICRAF	International Centre for Research in Agroforestry
IEK	Isotopic exchange kinetic
IFDC	International Centre for Soil Fertility and Agricultural Development, formerly International Fertilizer Development Center
IMPHOS	Institut Mondial du Phosphate
K	Potassium
L	Labile
LDCs	Least developing countries
LOI	Loss on ignition
MAP	Mono-ammonium phosphate
Mg	Magnesium
Mn	Manganese
Mo	Molybdenum
N	Nitrogen
Na	Sodium
NAC	Neutral ammonium citrate
NGO	Non-governmental organization
NORAGRIC	Centre for Environment and International Development, Agricultural University of Norway
NPK	Compound nitrogen, phosphorus and potassium containing fertilizers
NPV	Net present value
O	Oxygen
P	Phosphorus (element)
P fertilizer	Phosphatic fertilizer
P_2O_5	Phosphorus pentoxide (phosphate)
PAPR	Partially acidulated phosphate rock
Pb	Lead
P_i	Iron-oxide-impregnated paper
PPCL	Pyrites, Phosphates and Chemicals Ltd. (India)
PR	Phosphate rock
PR/S	Phosphate rock elemental sulphur assemblage
PR-DSS	Phosphate rock decision-support system
PRs	Phosphate rocks, PR materials, PR products
PSM	Phosphate solubilizing micro-organisms

RAE	Relative agronomic effectiveness
RELARF	Red Latinoamericana de Roca Fosforica
RPR	Reactive phosphate rock
RR	Relative response
Se	Selenium
SEM	Scanning electron microscopy
SFET	Soil fertility enhancing technology
SFI	Soil Fertility Initiative
Si	Silicon
SSP	Single superphosphate
STPP	Sodium tri-polyphosphate
SV	Substitution value
SV50	Substitution values of TSP (50% maximum yield) for North Carolina PR
TPR	Tilemsi phosphate rock
TSP	Triple superphosphate
U	Uranium
USGS	United States Geological Survey
V	Vanadium
VAM	Vesicular-arbuscular mycorrhizae
WSP	Water-soluble phosphate
XRD	X-ray diffraction
Zn	Zinc

Executive summary

In order to ensure food security in developing countries, there is a need for the sustainable intensification of agricultural production systems towards supporting productivity gains and income generation. In this context, novel, soil-specific technologies will have to be developed, pilot tested and transferred to farmers in a relatively short time. Phosphorus (P) is an essential nutrient element for plants and animals. The appropriate and sound utilization of phosphate rocks (PRs) as P sources can contribute to sustainable agricultural intensification, particularly in developing countries endowed with PR resources.

PR is a general term used to describe phosphate-bearing minerals. PR is a finite, non-renewable natural resource. Geological deposits of different origin are found throughout the world. Currently, few PR deposits are mined, and about 90 percent of world PR production is utilized by the fertilizer industry to manufacture P fertilizers, with the remainder being used to manufacture of animal feeds, detergents and chemicals. However, many PR deposits located in the tropics and subtropics have not been developed. One reason is that the characteristics of these PRs, though suitable for direct application, do not meet the quality standards required for producing water-soluble P (WSP) fertilizers using conventional industrial processing technology. Another reason is that the deposits are too small to warrant the investment needed for mining and processing. On a worldwide basis, there is ample supply of high-quality PRs for chemical processing and direct application for the foreseeable future.

PR is the primary raw material for producing P fertilizers. The phosphate compound in PRs is some form of the mineral apatite. Depending on the origin of the PR deposit and its geological history, the apatites may have widely differing chemical and crystallographic characteristics and physical properties. Distinctive suites of accessory minerals are also associated with PRs of various origins and geological histories. It is imperative to establish simple procedures for the standard characterization of PR sources, define quality standards for their direct application and categorize them. Well-known PR sources may be adopted as reference standards for comparison purposes.

The mineralogical, chemical and textural characteristics of phosphate ores and concentrates determine: (i) their suitability for various types of beneficiation processes to upgrade the ores and remove impurities; (ii) their adaptability for chemical processing by various routes; and (iii) their suitability for use as direct application phosphate rock (DAPR). The most important factors in the assessment for direct application are: grade, suitability for beneficiation, and the reactivity of the apatite. A comprehensive characterization matrix based on the integration of all the data obtained by various analytical methods brings out the beneficiation potential and probable best uses for a PR in fertilizer processing and as a direct application fertilizer.

There are various methods for evaluating PRs for direct application. The first approach uses empirical solubility tests with the PR being dissolved in chemical extracting solutions. The most common solutions are neutral ammonium citrate, 2-percent citric acid, and 2-percent formic acid, with the last option being preferred. The particle size of the PR and associated minerals in the PR can influence the result of the solubility test. Radioisotopic techniques can be used but they require skilled staff and special laboratory facilities. Soil incubation studies to evaluate

PRs are relatively simple. However, closed incubation studies have limitations as reaction products are not removed and the results would have limited utility unless used as short-term observations. Greenhouse experiments are valuable in that they enable the performance of the PR to be evaluated under controlled conditions. However, the final evaluation of the agronomic effectiveness of PR sources in a network of field experiments conducted over a number of growing seasons at representative sites in the agro-ecological regions of potential interest is essential. A series of recommendations for undertaking such evaluation have been developed. Such evaluation is also required for the assessment of the economic potential of PR deposits for domestic consumption.

Factors that influence the agronomic effectiveness of PRs are: the reactivity of the PR, soil properties, climate conditions, crop species and management practices. High carbonate substitution for phosphate in the apatite crystal structure, low content of calcium carbonate as accessory mineral and fine particle size (less than 0.15 mm) enhance the reactivity of PRs and their agronomic effectiveness. Rapid chemical tests are available to measure the reactivity of PRs. Increasing soil acidity, high cation exchange capacity, low levels of calcium (Ca) and phosphate in solution and high organic matter content favour PR dissolution. High phosphate retention capacity of soils may facilitate dissolution of PR but the availability of dissolved P will depend on the concentration of phosphate maintained in solution. Soils of medium phosphate status are considered to be more suitable for PR application than severely phosphate-deficient soils. Increasing rainfall invariably results in improved agronomic effectiveness of PRs. High agronomic effectiveness of PRs may be realized with perennial and plantation crops and also with legumes. To achieve maximum agronomic effectiveness, the PRs should be incorporated into the soil. While only a maintenance rate of P application need be applied as PR where the soil phosphate status is medium or above, very high rates of application are required for severely phosphate-deficient soils. Application time can be nearer to planting in very acid soils (pH < 5.5) and 4–8 weeks before planting in less acid soils.

The potential for using local PRs for direct application varies for each country as the complexity of the interactions occurring between specific local factors is evident in the tropical and subtropical world. Countries with substantial reactive PR reserves, such as Mali, Madagascar and Indonesia, offer considerable scope for the direct application of PRs. In countries endowed with less reactive PR reserves, such as Venezuela and Brazil, it is feasible to modify the PRs in order to improve their performance. In some cases, the economic advantage to the farmer can be considerable, as evidenced in the example on Venezuela. The Indian case study reveals import replacement as the major economic benefit resulting from the use of local Mussoorie PR. The New Zealand case study illustrates a success story on the use of reactive phosphate rocks (RPRs) for pastoral agriculture in Oceania.

Conventional soil tests for available P estimation are used for formulating fertilizer recommendations for the application of WSP sources. However, these common soil tests can underestimate (e.g. Bray I and Olsen) or overestimate (e.g. Bray II and Mehlich I) P availability in soils fertilized with water-insoluble PR. Separate calibration curves are needed for these two types of P fertilizer. Thus, the issue of developing suitable soil P testing methods for PR application brings a new dimension to phosphate research. There is need to develop appropriate soil tests that reflect P availability closely across a wide range of soil properties, PR sources and crop genotypes. Furthermore, soil tests should be suitable for both PRs and WSP fertilizers. Two recently developed soil tests show promise in soils fertilized with WSP and with PR sources. These are the iron-oxide-impregnated filter-paper strip test and the mixed cation-anion exchange resin membrane test. In both cases, the P extracted simulates P sorption by plant roots without the chemical reaction involved in the conventional tests. Available P measured in soils treated with both PR and WSP sources has shown: (i) good correlation with plant response; and (ii) both PR and WSP sources follow the same calibration curve. Further field trials are needed in order to test their suitability for developing recommendations for DAPR.

In order to provide sound guidelines for DAPR, it is essential to predict their agronomic effectiveness, crop yield increases and profitability. Several factors affect PR dissolution in soil under a specified set of conditions. DAPR as a P fertilizer is a more complex issue that involves consideration of a number of factors and their interactions. The development of a decision-support system (DSS) is the most effective approach for integrating all these factors and for providing an effective means of transferring research results to extension services. DSSs have been constructed to utilize available information to predict whether a given PR will be effective in a given crop environment. This publication discusses different types of DSSs, including the approaches adopted to develop a PR-DSS in New Zealand and Australia. It describes joint efforts currently underway by the International Centre for Soil Fertility and Agricultural Development (IFDC) and the Joint FAO/IAEA Division of Nuclear Techniques in Food and Agriculture (FAO/IAEA) to develop a more global DSS for use in tropical and subtropical countries for a range of food crops. Moreover, the PR-DSS could be linked to the phosphate model of the Decision Support System for Agrotechnology Transfer (DSSAT) family to predict crop yields, and GIS-based maps of crop response to PRs could also be compiled.

Phosphate-bearing minerals show a complex structure as a result of their geological origin and weathering processes. Their mineralogical and chemical composition is extremely variable. DAPR studies have focused mainly on the use of PR as a P source for crop production in acid soils, while limited research has been conducted on the effect of other beneficial and hazardous elements associated with PR use. Available information has suggested that PRs may also have potential agronomic value by provide some secondary nutrients, such as Ca and magnesium, and micronutrients, such as zinc and molybdenum, for plant growth depending on the type of PR sources. Reactive PRs containing free carbonates (calcite and dolomite) have been shown to reduce aluminium (Al) saturation of acid soils by raising soil pH and hence decreasing Al toxicity to plants. Limited studies also suggest that plant uptake of toxic heavy metals, notably cadmium, from PR is significantly lower than from WSP fertilizers produced from the same PR. More research is needed to investigate secondary nutrients, micronutrients, liming effect, and hazardous elements associated with PR use.

Not all PR sources are suitable for direct application. However, it is possible to utilize several means to improve their agronomic effectiveness under a particular set of condition. Selecting the appropriate process requires a good understanding of the factors hindering the agronomic effectiveness. The biological means (e.g. phospho-composting, inoculation with vesicular-arbuscular mycorrhizae, use of phosphate solubilizing micro-organisms, and use of P-efficient plant genotypes) are based on the production of organic acids to enhance PR dissolution and P availability to plants, and they hold good promise. The use of chemical means to produce partially acidulated PR (PAPR) is the most effective way of increasing the agronomic effectiveness of PRs and also saves energy. However, the production of PAPRs still requires fertilizer manufacturing plants. The physical means of dry mixing PR with a defined proportion of WSP fertilizer is promising and cost-effective. This simple approach should be promoted and evaluated under local soil and climate conditions.

Fertilizer legislation exists in many countries, especially developing ones. Its aim is to ensure that the fertilizer qualities meet the specifications set by the government in order to protect consumers' interests. As the bulk of PR production is utilized for WSP fertilizer manufacturing, only limited information on the legislation for DAPR is available. The agronomic effectiveness of PR for direct application depends on several factors and their specific interactions. Regulations for DAPR should consider three main factors: PR reactivity (solubility), soil properties (mainly soil pH) and crop species. Current legislation on PR for direct application considers only the quality of the PR, namely: total phosphorus pentoxide (P_2O_5) content, particle size distribution, and solubility. Based on recent research findings, this publication proposes several guidelines relating to issues on PR solubility and their interactions in a complex soil-plant system and concerning the establishment and revision of existing legislation.

In addition to technical considerations, a number of socio-economic factors and policy issues will determine PR production, distribution, adoption and use by the farmers. In principle, the use of PR sources should be promoted in countries where they are indigenously available. In countries with no PR deposits, it would be advisable to have a wide range of PRs available on the market in order to encourage price competition.

In conclusion, the appropriate and sound direct application of PR can contribute significantly towards sustainable agricultural intensification using natural plant nutrient resources. Although there has recently been substantial progress in scientific knowledge and technological developments on DAPR, further exploration of the specific issues is imperative.

Chapter 1
Introduction

PHOSPHORUS IN THE SOIL-PLANT SYSTEM

Phosphorus (P) is an element that is widely distributed in nature and occurs, together with nitrogen (N) and potassium (K), as a primary constituent of plant and animal life. P plays a series of functions in the plant metabolism and is one of the essential nutrients required for plant growth and development. It has functions of a structural nature in macromolecules such as nucleic acids and of energy transfer in metabolic pathways of biosynthesis and degradation. Unlike nitrate and sulphate, phosphate is not reduced in plants but remains in its highest oxidized form (Marschner, 1993).

P is absorbed mainly during the vegetative growth and, thereafter, most of the absorbed P is re-translocated into fruits and seeds during reproductive stages. P-deficient plants exhibit retarded growth (reduced cell and leaf expansion, respiration and photosynthesis), and often a dark green colour (higher chlorophyll concentration) and reddish coloration (enhanced anthocyanin formation). It has been reported that the level of P supply during reproductive stages regulates the partitioning of photosynthates between the source leaves and the reproductive organs, this effect being essential for N-fixing legumes (Marschner, 1993). Healthy animals and human beings also require adequate amounts of P in their food for normal metabolic processes (FAO, 1984, 1995a).

This nutrient is absorbed by plants from the soil solution as monovalent (H_2PO_4) and divalent (HPO_4) orthophosphate anions, each representing 50 percent of total P at a nearly neutral pH (pH 6–7). At pH 4–6, H_2PO_4 is about 100 percent of total P in solution. At pH 8, H_2PO_4 represents 20 percent and HPO_4 80 percent of total P (Black, 1968).

The physico-chemistry of P in most mineral soils is rather complex owing to the occurrence of series of instantaneous and simultaneous reactions such as dissolution, precipitation, sorption and oxidation-reduction. The P-soluble compounds have very high reactivity, low solubility indices and low mobility. Mineralization and immobilization of organic P compounds are relevant processes for P cycling in soils containing significant amounts of organic matter (Black, 1968; FAO, 1984).

Where a water-soluble P (WSP) fertilizer is applied to the soil, it reacts rapidly with the soil compounds. The resulting products are sparingly soluble P compounds and P adsorbed on soil colloidal particles (FAO, 1984). A low P concentration in the soil solution is usually adequate for normal plant growth. For example, Fox and Kamprath (1970) and Barber (1995) suggested that 0.2 ppm P was adequate for optimum growth. However, for plants to absorb the total amounts of P required to produce good yields, the P concentration of the soil solution in contact with the roots requires continuous renewal during the growth cycle.

Under continuous cultivation, P inputs, in particular water-soluble fertilizers, must be added to either maintain the soil P status of fertile soils or increase that of soils with inherent low P fertility. Therefore, soil, crop, water, P-fertilizer management practices, climate conditions, etc. are important factors to be considered when attempting to formulate sound P-fertilizer recommendations and obtain adequate crop yield responses (FAO, 1984, 1995a).

THE NEED FOR SUSTAINABLE DEVELOPMENT

The present world population of 6 000 million is expected to reach 8 000 million by 2020 and 9 400 million by 2050. By then, the population of the developing world will probably be 8 200 million (Lal, 2000). Approximately 50 percent of the potentially arable land is currently under arable and permanent crops. A further 2 000 million ha has been degraded and land degradation continues through a wide range of processes, mainly related to mismanagement by humankind (Oldeman, 1994; FAO, 1995b; UNEP, 2000).

Against this global background, several developing countries will face major challenges in achieving sustainable food security. This is because of their available per capita land area, severe scarcity of freshwater resources, particular socio-economic conditions of their agriculture sector, and internal structures and conflicts (Hulse, 1995).

Enhancing sustainable food production will require a proper use of the available land and water resources, i.e.: (i) agricultural intensification on the best arable land; (ii) adequate utilization of marginal lands; and (iii) prevention and restoration of soil degradation.

In order to increase the intensification, diversification and specialization of agricultural production systems towards supporting productivity gains and income generation, innovative soil-specific technologies will have to be developed, pilot tested and transferred in a relatively short time. These technologies will address priority issues such as: (i) enhancing cropping intensity by exploiting genotype differences in adaptation to particular environments and nutrient use efficiency; (ii) increasing nutrient use efficiency and recycling through integrated nutrient sources management in cropping systems; (iii) soil and water conservation through crop residue management and conservation tillage; and (iv) improving water use efficiency through the development of efficient methods of irrigation, water harvesting and recycling (Lal, 2000).

In preventing and reversing soil degradation, the main sustainability issues will concern controlling soil erosion and associated sedimentation and risks of eutrophication of surface water and contamination of groundwater (UNEP, 2000). Similarly, enhancing soil carbon sequestration in cropland to improve soil quality and productivity and mitigate the greenhouse effect will also be a matter of concern (Lal, 1999).

The agricultural frontier is likely to expand into marginal lands with harsh environments that contain fragile soils with a lower productive capacity and a higher risk of degradation. The use of plant genotypes with adequate yield potential, efficient in nutrient use and tolerant to soil and environmental stresses (drought, acidity, salinity, frost, etc.) will be of strategic importance. Their use is becoming increasingly important in many international and national breeding programmes (Date *et al.*, 1995; Pessarakli, 1999). This approach is currently being utilized for the sustainable management of P-deficient acid soils (Rao *et al.*, 1999; Hocking *et al.*, 2000; IAEA, 2000; Keerthisinghe *et al.*, 2001).

The development and application of an integrated nutrient management approach in the agriculture of developing countries will imply the use of chemical fertilizers and natural sources of nutrients, such as phosphate rocks (PRs), biological nitrogen fixation (BNF), and animal and green manures, in combination with the recycling of crop residues (FAO, 1995a). The utilization of these technologies requires the assessment of the nutrient supply from the locally available materials applied as nutrient sources, their tailoring to specific cropping systems and the provision of guidelines for their application (FAO, 1998; Chalk *et al*., 2002). This is particularly the case of indigenous PR resources in the tropics.

ADDRESSING THE P CONSTRAINT IN TROPICAL ACID SOILS

Extensive tracts of land in the tropical and subtropical regions of Asia, Africa and Latin America contain highly weathered and inherently infertile soils. These areas generate low crop yields and are prone to land degradation as a result of deforestation, overgrazing and inadequate farming practices. In addition to socio-economic factors, the main constraints are soil acidity and low inherent N and P fertility (Lal, 1990; Formoso, 1999). While N inputs can be obtained from sources such as BNF, crop residues and other organic sources, P inputs need to be applied in order to improve the soil P status and ensure normal plant growth and adequate yields. Tropical and subtropical soils are predominantly acidic and often extremely P deficient with high P-sorption (fixation) capacities. Therefore, substantial P inputs are required for optimum growth and adequate food and fibre production (Sanchez and Buol, 1975; Date *et al*., 1995).

Manufactured WSP fertilizers such as superphosphates are commonly recommended for correcting P deficiencies. However, most developing countries import these fertilizers, which are often in limited supply and represent a major outlay for resource-poor farmers. In addition, intensification of agricultural production in these regions necessitates the addition of P inputs not only to increase crop production but also to improve soil P status in order to avoid further soil degradation. Therefore, it is imperative to explore alternative P inputs. In this context, under certain soil and climate conditions, the direct application of PRs is an agronomic and economically sound alternative to the more expensive superphosphates in the tropics (Chien and Hammond, 1978; Truong *et al*., 1978; Zapata *et al*., 1986; Hammond *et al*., 1986b; Chien and Hammond, 1989; Chien *et al*., 1990b; Sale and Mokwunye, 1993).

USE OF PHOSPHATE ROCKS IN INDUSTRY AND AGRICULTURE

Phosphate rock denotes the product obtained from the mining and subsequent metallurgical processing of P-bearing ores. In addition to the main phosphate-bearing mineral, PR deposits also contain accessory or gangue minerals and impurities. Although considerable amounts of accessory minerals and impurities are removed during beneficiation, the beneficiated ore still contains some of the original impurities. Such impurities include silica, clay minerals, calcite, dolomite, and hydrated oxides of iron (Fe) and aluminium (Al) in various combinations and concentrations, some of which may have a marked influence on the performance of a PR used for direct application (UNIDO and IFDC, 1998). Thus, currently PR is the trade name of about 300 phosphates of different qualities in the world (Hammond and Day, 1992).

PRs can be used either as raw materials in the industrial manufacture of WSP fertilizers or as P sources for direct application in agriculture.

Phosphate rocks as raw materials for P-fertilizer manufacturing

The world phosphate industry is based on the commercial exploitation of some PR deposits. In spite of their extremely variable composition, PRs are the commercial source of P used as the raw material for manufacturing P fertilizers and certain other chemicals. Unlike other vital commodities, such as Fe, copper (Cu) and sulphur (S), there is little opportunity for substitution or recycling. Phosphate ranks second (coal and hydrocarbons excluded) in terms of gross tonnage and volume of international trade.

The fertilizer industry consumes about 90 percent of world PR production. Sulphuric acid and PR are the raw materials used in the production of single superphosphate (SSP) and phosphoric acid. Phosphoric acid is an important intermediate by-product that is used to make triple superphosphate (TSP) and ammonium phosphate. Highly concentrated, compound NPK formulations now form the mainstay of the world fertilizer industry (Engelstad and Hellums, 1993; UNIDO and IFDC, 1998).

PR is also used for industrial purposes and for the production of animal feed supplements and food products. Another important use is in the manufacture of elemental P and its derivatives, in particular sodium tri-polyphosphate, a major component of heavy-duty laundry detergents (Hammond and Day, 1992; UNIDO and IFDC, 1998).

PR deposits are widely distributed throughout the world, both geographically and geologically, and there are very large resources capable of meeting anticipated demand for the foreseeable future. Estimates generally indicate a total of 200 000–300 000 million tonnes of PR of all grades. Of these total estimates, a large proportion includes deposits composed of carbonate-rich PR, whose commercial exploitation depends either on the development of new beneficiation technology or on changes in economic conditions (British Sulphur Corporation Limited, 1987; Notholt et al., 1989).

About 80 percent of world PR production is derived from deposits of sedimentary marine origin, some 17 percent is derived from igneous rocks and their weathering derivatives, and the remainder comes from residual sedimentary and guano-type deposits.

Sedimentary PRs are mainly composed of apatites. These apatites exhibit extensive isomorphic substitution in the crystal lattice. Thus, they have a wide variation in chemical composition and accordingly show a wide range of properties. In sedimentary deposits, the main phosphate minerals are francolites (microcrystalline carbonate fluorapatites), which occur in association with a variety of accessory minerals and impurities (McClellan and Van Kauwenbergh, 1990a).

The phosphate content or grade of PR is conventionally expressed as phosphorus pentoxide (P_2O_5). In some low-grade commercial deposits, this may be as low as 4 percent. In the phosphate industry, the phosphate content of the rock is usually expressed as tricalcium phosphate and traditionally referred to as bone phosphate of lime (P_2O_5 x 2.1853 = BPL). The latter term is reminiscent of the time when bones were the principal source of phosphate in the fertilizer industry. Manufacturers of phosphoric acid and P fertilizers normally stipulate a minimum content of 28 percent P_2O_5, and most marketed grades of PR contain more than 30 percent P_2O_5 (65 percent BPL). To meet this requirement, most phosphate ores undergo beneficiation by washing and screening, de-liming, magnetic separation and flotation (Hammond and Day, 1992; UNIDO and IFDC, 1998).

Phosphate rock for direct application in agriculture

As mentioned above, PRs mainly of sedimentary origin are suitable for direct application because they consist of fairly open, loosely consolidated aggregates of microcrystals with a relatively large specific surface area. They show a considerable proportion of isomorphic substitution in the crystal lattice and contain a variable proportion and amounts of accessory minerals and impurities. Thus, it is reported that these PRs are suitable for direct application to soils under certain conditions (Khasawneh and Doll, 1978; Chien, 1992; Chien and Friesen, 1992; Chien and Van Kauwenbergh, 1992; Chien and Menon, 1995b; Rajan *et al.*, 1996; Zapata, 2003).

The practice of direct application of PR sources as fertilizers has several advantages:

- PRs are natural minerals requiring minimum metallurgical processing. The direct application of PRs avoids the traditional wet acidification process for WSP fertilizers and circumvents the production cycle of polluting wastes such as phospho-gypsum and greenhouse gases, thus resulting in energy conservation and protection of environment from industrial pollution.

- Being natural compounds, PRs can be used in organic agriculture.

- Direct application enables the utilization of PR sources that cannot be utilized for industrial purposes in the manufacture of WSP fertilizers and phosphoric acid.

- PRs suitable for direct application (reactive) can be more efficient than WSP fertilizers in terms of P recovery by plants under certain conditions.

- Based on the unit cost of P, natural or indigenous PR is usually the cheapest.

- Because of their extremely variable and complex chemical composition, PRs are sources of several nutrients other than P. They are usually applied to replenish the soil P status, but upon dissolution they also provide other nutrients present in the PR. Application of medium and highly reactive PRs to highly weathered tropical acid soils has a potential trigger effect on plant growth and crop yields as a result not only of P release but also of their effects on increasing exchangeable calcium (Ca) and reducing Al saturation. The resulting harvest products and residues have a better nutritional quality (higher P content than unfertilized plants). The incorporation of such organic residues enhances biological activity and soil carbon (C) accumulation, leading to improved soil physical and chemical properties. Thus, PRs have an important role in contributing to improving soil fertility and soil degradation control, in particular nutrient mining (depletion).

However, this practice also has some limitations:

- Not all PRs are suitable for direct application. The effectiveness of some medium-to-low reactive PRs needs to be enhanced by biological and physico-chemical processes. Specific technologies must be developed and tested and their economics must be assessed on a case-by-case basis.

- Not all soils and cropping systems are suitable for PRs of different origin. A standardized characterization of PRs is required as are guidelines for providing them.

- There is a lack of knowledge on the main factors and conditions affecting the agronomic effectiveness of PRs and an inability to predict their effectiveness as well as a lack of assessment of socio-economic factors, financial benefits and government policies. In this respect, progress is being made in the development of decision-support systems (DSSs) for integrating all the factors that influence the use and adoption of PR technology.

- The low grade of some PRs compared with high-grade commercial P fertilizers makes them more expensive at the point of application. This economic evaluation is very dynamic and it should be made at the time of exploitation of the PR deposit.

- Sedimentary PRs show a very complex structure as a result of their different origin in nature and even within a particular geological deposit. Thus, they have extremely variable chemical constituents and may contain elements such as heavy metals and even radionuclides that upon dissolution of the PR in the soil may be harmful at some concentrations.

Several PR research projects have recently made considerable progress in addressing the issues mentioned above.

Historical account of the use of phosphate rocks in agriculture

The direct application of ground, natural PR as a source of P for crops is a practice that has been utilized with varying degrees of popularity over the years. Numerous field and greenhouse experiments have been conducted during the past 100 years or more to assess the capabilities of these materials to supply P to crops and to define the most favourable conditions for their application. The results obtained have been reported as erratic and sometimes conflicting, leading to confusion and disagreement on the utilization of PRs (Khasawneh and Doll, 1978).

In general, experiments conducted in the past showed that PRs were most effective when applied to plantation crops in acid soils of the tropics. Moreover, under conditions where suboptimal yields were expected because of limits on additional inputs, i.e. extensive pasture systems, local PRs were thought to be a possible suitable source of P. However, no conclusive evidence could be obtained either for or against their adoption in the majority of cases.

The main reason for this situation was the lack of understanding of the various factors affecting the agronomic effectiveness of PRs. Since then, significant progress has been made on the evaluation of the main factors affecting their agronomic effectiveness. A milestone work on this topic was the comprehensive review by Khasawneh and Doll (1998). They examined the influence of the inherent PR factors (mineralogy, chemical composition, solubility tests and physical properties), soil factors (pH, soil texture, soil organic matter, soil P status, available P, P fixation, Ca content, etc.) and plant factors (growth cycle, P demand and pattern of P uptake, root system, rhizosphere properties, etc.).

Hammond *et al.* (1986b) conducted an extensive review on the agronomic value of indigenous PR sources located in the tropics, highlighting the potential use of partially acidulated products (PAPR) throughout Latin America, Africa and Asia. More recently, an updated review focusing on the fundamentals of PR dissolution in soils, the concepts of agronomic effectiveness of PRs and approaches to assessing the economics of PR utilization was conducted by Rajan *et al.* (1996).

National projects on the use of PRs for pastures in different environments have been implemented for more than 50 years in New Zealand and Australia (Bolan *et al.*, 1990; Rajan, 1991a, 1991b; Bolland *et al.*, 1997). Field trials with reactive PRs (RPRs) were carried out in New Zealand in the early 1980s (Hedley and Bolan, 1997; 2003), and in the period 1991–96 in Australia with implementation of the National Reactive PR Project (Simpson *et al.*, 1997; Sale *et al.*, 1997a). In both countries, significant progress was made in determining the soil, climate and pasture conditions under which RPR products are effective substitutes for WSP fertilizers and in defining the level of reactivity required for RPRs to be effective.

Regional networks in Latin America (Red Latinoamericana de Roca Fosforica – RELARF) and in Asia (East and Southeast Asia Program of the Potash and Phosphate Institute of Canada) have also been operational for some time in the tropics and subtropics of these regions. They have held periodical meetings to report on their research results (Dahayanake *et al.*, 1995; Hellums, 1995a; Johnston and Syers, 1996; Casanova and Lopez Perez, 1991; Casanova, 1995, 1998; RELARF, 1996; Zapata *et al.*, 1994; Besoain *et al.*, 1999). A number of studies have also been conducted in almost all countries of Africa, but the results are scattered in many individual reports of limited circulation. Truong *et al.* (1978) carried out a comprehensive study with several PR sources from West Africa. Proceedings of regional meetings on the use of fertilizers and local mineral resources for sustainable agriculture in Africa have been published (Mokwunye and Vlek, 1986; Gerner and Mokwunye, 1995). Reports synthesizing results from PR studies have recently been produced. A report by FAO (2001b) presents the results from field agronomic trials in West Africa while Appleton (2001) has provided a comprehensive review of local phosphate resources in sub-Saharan Africa in the context of sustainable development. In addition, several international meetings organized periodically by the Institut Mondiale du Phosphate (IMPHOS) have provided an international forum for reporting and exchanging information on phosphate research (IMPHOS, 1983, 1992).

In the 1970s, the International Centre for Soil Fertility and Agricultural Development (IFDC) began to conduct research with a focus on the use of indigenous PR deposits as a source of P for crop production in developing countries. This PR research has been an important component of the IFDC's programme for many years (Chien and Hammond, 1978, 1989; Chien *et al.*, 1987b; Chien and Friesen, 1992; Hellums *et al.*, 1990; Hellums, 1992; Chien, 1995; Chien and Menon, 1995b). This research has also included the development and evaluation of modified PR products as well their economic assessment (Hammond *et al.*, 1986b; Menon and Chien, 1990, 1996; Hellums *et al.*, 1992; Chien and Menon, 1995a; Baanante, 1998; Baanante and Hellums, 1998; Henao and Baanante, 1999). In the 1990s, the IFDC commenced research on environmental issues associated with PRs and P fertilizers because they contain varying amounts of potentially hazardous elements such as heavy metals (Hellums, 1995b; Iretskaya *et al.*, 1998; Iretskaya and Chien, 1999).

In 1994, the World Bank, in consultation with centres operated by the Consultative Group on International Agricultural Research (CGIAR) and university research groups from industrialized countries, launched the initiative Development of National Strategies for Soil Fertility Recapitalization in sub-Saharan Africa (World Bank, 1994; Valencia *et al.*, 1994; Buresh *et al.*, 1997; Baanante, 1998). One of the outcomes was the document Framework for National Soil Fertility Improvement Action Plans. This initiative included the use of PR as a capital investment in natural resources in Africa, where the situation is paradoxical because the soils are extremely poor in P in spite of the existence of numerous PR deposits. It was postulated that a one-time massive application of PR rock would overcome the soil P-sorption capacity problem and replenish the soil P capital. Further small applications of water-soluble fertilizers would be then more available to crops and of increased efficiency. The World Bank mandated several research organizations to conduct case studies in Burkina Faso, Madagascar and Zimbabwe. Although the results from these case studies suggested that the available PR resources in these countries could be used as capital investment to replenish soil P status, they were not conclusive owing to the lack of a comprehensive evaluation of the factors that influence the use and adoption of PR technology in each country (World Bank, 1997).

Within the framework of the Integrated Plant Nutrition Systems promoted by the Land and Water Development Division (AGL) of FAO and the national action plans of the Soil Fertility Initiative (SFI) for sub-Saharan countries, PRs are considered as important potential locally available P inputs that can be used gainfully (FAO, 2001a). The AGL has instituted several studies on the agro-economic assessment of PRs for direct application in selected countries. Results of practical utility and policy guidelines can be drawn from these and other studies.

In the period 1993–99, the Joint FAO/IAEA Division of Nuclear Techniques in Food and Agriculture implemented a research network comprising 21 institutions of developing and developed countries with the aim of evaluating the agronomic effectiveness of P fertilizers in particular PRs using nuclear and related techniques. The data obtained have demonstrated the potential of PRs to improve soil fertility and increase agricultural production under certain conditions. The results have been published in IAEA documents and scientific journals (Zapata, 1995, 2000, 2002, 2003; IAEA, 2000, 2002). Important follow-up activities of this project are a joint FAO/IAEA-IFDC undertaking to develop a DSS for the direct application of PRs and the preparation of a Web site and technical publications for the wider dissemination of results to professional technical staff, and decision-makers, including extension specialists and progressive farmers (Chien *et al.*, 1999; Heng, 2000, 2003; Singh *et al.*, 2003).

In view of recent developments and practical experiences on technologies for the direct application of PR and related appropriate technology, the IFDC, in collaboration with the Malaysian Society of Soil Science, the Potash and Phosphate Institute and the Potash and Phosphate Institute of Canada – East and Southeast Asia Program, organized an international meeting in Kuala Lumpur. The event attracted more than 100 participants from more than 30 countries from all over the world representing different national and international research networks on PR, producers, dealers and users of PR for direct application. The latest agronomic research results on the use of natural PRs and modified products as influenced by sources of PR, types of soil, management practices and cropping systems were reviewed, and updated information on the production and agronomic use of PR was gathered from the PR producers, dealers and users. The event also provided an international forum for discussing future trends on the use of indigenous or imported PRs for direct application to increase crop production and lower production costs (IFDC, 2003).

In conclusion, extensive research on the agronomic potential and actual effectiveness of PRs as sources of P has been carried out in Africa, Asia, Latin America and elsewhere. A wealth of information is available but it is scattered in several publications of meetings, technical reports, scientific and other publications. Overall, information on DAPR is limited and there remain areas and topics associated with DAPR where further attention is needed.

The bulletin

From the sections above, it may be also inferred that there is a need for a comprehensive publication to cover the key topics dealing with the utilization of PRs in agriculture, including the latest information on PR research, and to provide guidelines for the direct application of PR to the acid soils of the tropics and subtropics. This bulletin is an attempt to meet this need. It is conceived as a technically oriented document for a target audience comprising policy- and decision-makers, the scientific community, higher level extension workers, non-governmental organizations (NGOs) and other stakeholders involved in sustainable agricultural development at local, national, regional and international levels.

The chapters in this bulletin provide an overview of the scientific basis for PR use, and they present technical information on the most relevant issues related to the use of PR sources for direct application. They provide: comprehensive coverage of world PR deposits; characterization of PR sources; evaluation methodologies of PR sources for direct application; analysis of the biophysical and farming factors that affect the agronomic effectiveness of PR sources; and also analysis of the socio-economic conditions and other factors that ultimately influence the use and adoption of PR technologies as a capital investment to trigger agricultural intensification. The chapters also cover: development and use of DSSs for DAPR; soil P testing for PR application; available technologies for enhancing the agronomic effectiveness of indigenous PR sources;

environmental issues; and legislation guidelines. Finally, in the light of current knowledge and available technologies, future researchable areas and priorities are defined in an epilogue. The bibliography section provides a comprehensive list of references.

Chapter 2
World phosphate deposits

Phosphate rock (PR) is a general term that describes naturally occurring mineral assemblages containing a high concentration of phosphate minerals. The term refers to both unbeneficiated phosphate ores and concentrated products. Sedimentary deposits have provided about 80–90 percent of world production in the last ten years. They occur in formations of widely varying geological ages, exhibit a range of chemical compositions and physical forms, often occur as relatively flat-lying thick beds, and may underlie shallow overburden. The deposits that account for most of world PR production are in Morocco and other African countries, the United States of America, the Near East and China. Most sedimentary deposits contain the carbonate-fluorapatite called francolite (McConnell, 1938). Francolites with high carbonate for phosphate substitution are the most highly reactive and are the most suitable for direct application as fertilizers or soil amendments (Chapter 3).

Igneous deposits have provided about 10–20 percent of world production in the last ten years. They are exploited in the Russian Federation, Canada, South Africa, Brazil, Finland and Zimbabwe but also occur in Uganda, Malawi, Sri Lanka and several other locations. These deposits usually contain varieties of fluorapatite that are relatively unreactive and are the least suitable for direct application. The weathering products of igneous and sedimentary apatites (iron and aluminium phosphate minerals) are generally not useful for direct application in agriculture in their natural state.

Phosphate is the component of agronomic interest in these rocks. The higher the phosphate (P_2O_5) content as apatite, the greater the economic potential of the rock. Factors that are important in the chemical conversion of PRs to fertilizer (free carbonates, iron (Fe), aluminium (Al), magnesium (Mg) and chloride) are often not important where the rock is to be used for direct application (Gremillion and McClellan, 1975; McClellan and Gremillion, 1980; Van Kauwenbergh and Hellums, 1995).

WORLD PHOSPHATE ROCK PRODUCTION

Table 1 lists world PR production for 1999, the most recent year for which firm figures are available. The main four producers of PR (the United States of America, China, Morocco and Western Sahara, and the Russian Federation) produced about 72.0 percent of the world total. The leading 12 producers accounted for more than 93 percent of the world total. Twenty other countries produced the remaining 6–7 percent of world production.

On a worldwide basis, the production and consumption of direct application phosphate rock (DAPR) are very limited, and reliable data are often difficult to obtain and evaluate. Many countries do not classify DAPR as a fertilizer and consumption statistics may not include it. Information may have to be obtained through unofficial channels, and its quality may be highly variable. It may be necessary to estimate DAPR by subtracting the amount of PR used for other purposes from the total amount of PR imported or used within a country. The figures presented in this chapter for DAPR are indicative rather than firm figures.

TABLE 1
World phosphate rock production, 1999

	Product (1 000 tonnes)	World total %
United States of America	40 867	28.1
China	30 754	21.1
Morocco and Western Sahara	21 986	15.1
Russian Federation	11 219	7.7
Subtotal top four	**104 826**	**72.0**
Tunisia	8 006	5.5
Jordan	6 014	4.1
Brazil	4 301	2.9
Israel	4 128	2.8
South Africa	2 941	2.0
Syrian Arab Republic	2 084	1.4
Senegal	1 879	1.3
Togo	1 715	1.2
Subtotal top twelve	**135 894**	**93.4**
India	1 623	1.1
Algeria	1 093	0.8
Egypt	1 018	0.7
Mexico	955	0.7
Kazakhstan	900	0.6
Finland	734	0.5
Viet Nam	710	0.5
Christmas Island	683	0.5
Nauru	604	0.4
Iraq	415	0.3
Venezuela	366	0.3
Canada	350	0.3
Australia	145	0.1
Uzbekistan	139	< 0.1
Zimbabwe	124	< 0.1
Democratic People's Republic of Korea	70	< 0.1
Sri Lanka	30	< 0.1
Peru	15	< 0.1
Colombia	4	< 0.1
World total	**145 472**	**100.0**

Source: Mew, 2000.

TABLE 2
World consumption of direct application phosphate rock

	World phosphate consumption (nutrient basis) %	Approximate tonnage (million tonnes at 30% P_2O_5)
1975	5.6	4.8
1980	4.9	5.2
1985	4.0	4.5
1990	3.0	3.6
1991	1.7	2.0
1995	1.5	1.5
1998	1.4	1.5

Source: Van Kauwenbergh, 2003.

One estimate of world DAPR consumption from 1975 to 1998 (Table 2) shows a fall in consumption from 5.6 to 1.4 percent of total P_2O_5 consumption. This is equivalent to about 1.5 million tonnes of product averaging about 30 percent P_2O_5. Another source (Maene, 2003) indicates a drop in consumption of DAPR from 1.66 million tonnes P_2O_5 in 1980 to 0.57 million tonnes P_2O_5 in 1998. This is equivalent to a consumption of about 1.9 million tonnes of product in 1998 at 30 percent P_2O_5. While the exact current amount of DAPR consumption is difficult to determine, it appears that world consumption is less than 2.0 million tonnes of product per year. In the former Soviet Union (FSU), consumption of DAPR decreased from about 900 000 tonnes P_2O_5 to about 350 000 tonnes in 1991; its use had previously been government mandated on collective farms. In China, DAPR consumption fell from about 300 000 tonnes P_2O_5 in the early to mid-1980s to less than 30 000 tonnes P_2O_5 in 1995 (Rong, 1995). In China, DAPR is in use only near mines in areas with acid soils.

PR from Tunisia, China, Christmas Island, Egypt, Israel, Jordan, Morocco and Peru continues to be produced and used for direct application. Colombia, India, Sri Lanka and Venezuela also produce DAPR. Sales of North Carolina DAPR were once approximately 250 000–300 000 tonnes/year; however, this marketing activity has been suspended and North Carolina PR is not currently available for direct application.

WORLD PHOSPHATE ROCK RESERVES AND RESOURCES

There is no accepted worldwide system for classifying PR reserves and resources. A system developed in the United States of America (U.S. Bureau of Mines and U.S. Geological Survey,

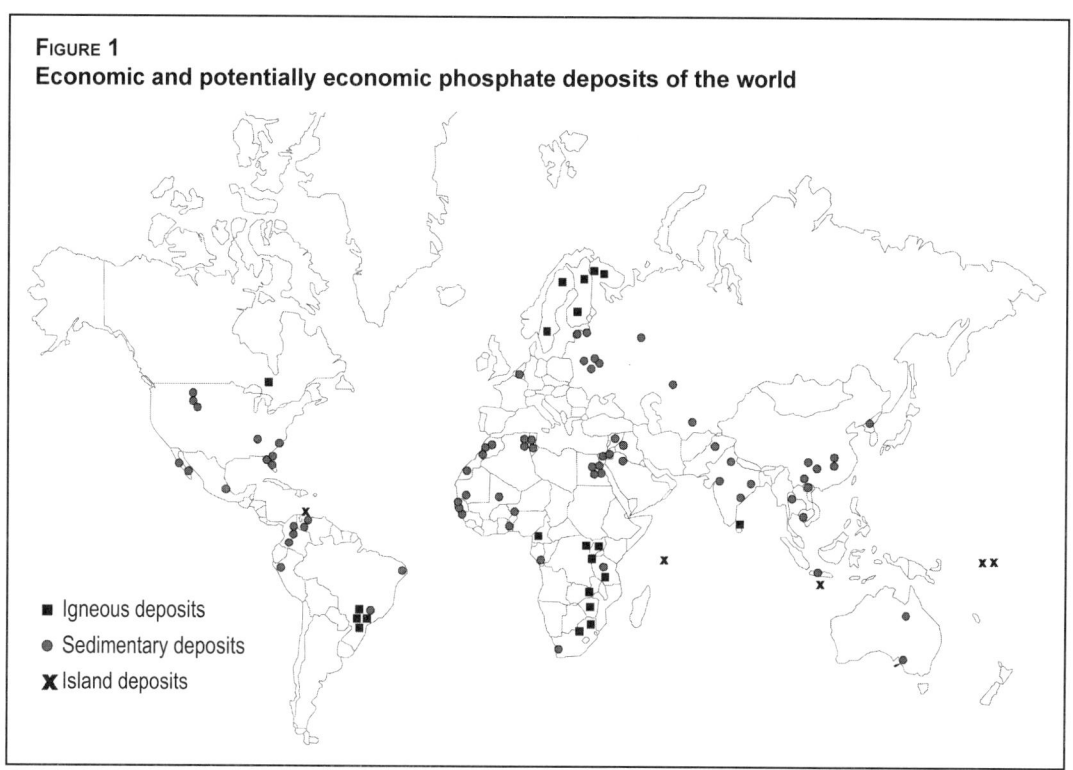

FIGURE 1
Economic and potentially economic phosphate deposits of the world

■ Igneous deposits
● Sedimentary deposits
✗ Island deposits

1981; U.S. Geological Survey, 1982) defines reserves as "identified resources of a mineral that can be extracted profitably with existing technology and under present economic conditions" (Brobst and Pratt, 1973). Reserve estimates may be stated as the total amount of minable rock in the ground or as the amount of recoverable product. Many authors do not distinguish between reserves and non-economic resources when reporting the size of deposits. Thus, substantial differences in reserve and/or resource estimates may exist between various sources. It is prudent to acknowledge that such discrepancies exist and that such figures should serve only as order-of-magnitude estimates.

Figure 1 shows a map of PR deposits currently being mined, those that have been mined in the recent past, and those that have been shown to be potentially economic. A tabulation (Table 3) of the ten main producing countries and their reserve bases shows that these countries possess about 90 percent of the world's phosphate reserves. Based on current extraction rates and economic conditions in the 1990s, more than half of these countries will have exceeded the life of their reserves in less than 20 years.

Sheldon (1987) categorized phosphate reserves and resources according to continents and regions (Table 4). At first glance, each main

TABLE 3
World phosphate rock reserves and reserve base

	Reserves[a]	Reserve base[b]
	(1 000 tonnes)	
United States of America	1 000 000	4 000 000
China	500 000	1 200 000
Israel	180 000	180 000
Jordan	900 000	1 700 000
Morocco and Western Sahara	5 700 000	21 000 000
Senegal	150 000	1 000 000
South Africa	1 500 000	2 500 000
Togo	30 000	60 000
Tunisia	100 000	600 000
Russian Federation	150 000	1 000 000
Other countries	1 200 000	4 000 000
World total	**12 000 000**	**37 000 000**

a. Cost less than US$40/tonne. Cost includes: capital, operating expenses, taxes, royalties, and a 15-percent return on investment f.o.b. mine.
b. Criteria for reserve base established by a joint U.S. Bureau of Mines and U.S. Geological Survey working group.

Source: US Bureau of Mines, 2001.

TABLE 4
World economic identified phosphate concentrate resources

Continent and country	Reserves	Inferred reserve base and reserve base
	(million tonnes)	
North America		
United States of America	1 260	22 587
Canada	37	37
Mexico	208	2 416
	1 505	25 040
South America		
Brazil	551	558
Peru	428	1 353
Other	120	211
	1 099	2 122
Africa		
Algeria	250	435
Morocco and Western Sahara	2 100	62 575
Senegal	155	200
South Africa	1 800	6 781
Tunisia	70	1 021
Other	82	1 216
	4 457	72 228
Asia		
China	170	1 000
Iraq	296	296
Israel	70	449
Jordan	530	1 542
Other	241	807
	1 307	4 094
Oceania		
Australia	340	850
Other	51	51
	391	901
Europe		
Former Soviet Union	6 500	8 000
Finland	-	46
	6 500	8 046
World total	15 259	112 431

Source: Sheldon, 1987.

continent/region has ample 'reserves' of phosphate, with the possible exception of Oceania (391 million tonnes). However, regionally, a few countries, or even one country, may dominate. In North America, the United States of America possesses 84 percent of the reserves. In Europe, the FSU countries possess 99 percent of the reserves. Within Africa, Morocco, South Africa, Algeria, Senegal and Tunisia possess 98 percent of the reserves. The South American reserves lie mainly in Brazil and Peru (97 percent). In Asia, 88 percent of the reserves occur in Iraq, Israel, Jordan and China. Australia possesses 87 percent of the reserves of Oceania.

As it is difficult to assign firm figures for current DAPR production and consumption, it is also difficult to estimate world resources of PR most suitable for direct application. The criterion that a high-solubility DAPR must have more than 5.9 percent P_2O_5 soluble in neutral ammonium citrate (NAC) (Hammond and Leon, 1983) limits the number of deposits to be considered. Using this criterion, the North Carolina, Peru and Tunisia deposits may be considered as containing the most significant world resources of the most highly reactive PR. Although there are a number of small deposits, or portions of deposits, that may contain PR with relatively high solubility, this analysis does not consider these deposits because resource estimates are not available for many of them.

Using the inferred reserve base and reserve base figures in (Table 4) for Peru and Tunisia and a figure of 730 million tonnes for the resources of the North Carolina deposits (Stowasser, 1991) (producible at less than US$60/tonne), the total resources of DAPR with an NAC solubility of more than 5.9 percent P_2O_5 amount to 3 100 million tonnes.

Using Sheldon's figure of a total world inferred reserve base of 112 431 million tonnes, world resources of high-solubility DAPR amount to about 2.8 percent of total world resources.

Considering the more conservative reserve base figure for the Tunisian deposits from the United States Geological Survey (USGS) (600 million tonnes) (Table 3), the resource estimates of Sheldon (1987) and Stowasser (1991) for Peru and North Carolina, respectively, and the total world reserve base of the USGS (37 000 million tonnes), world resources of high-solubility DAPR may amount to about 7.2 percent of total world resources.

Although detailed consideration of small deposits and portions of large deposits containing highly soluble PR or redefining the solubility level of the PR chosen to perform the analysis may serve to increase the amount of potential world resources of the most suitable DAPR, it is apparent that the potential world resources of the most suitable DAPRs are limited and that these represent only a fraction of total world PR resources.

FUTURE TRENDS IN WORLD PHOSPHATE ROCK PRODUCTION

Among the current leading four PR-producing countries, Morocco is in the most advantageous position and may possess over half of the world's phosphate reserves. In the last 12 years, China has approximately doubled its production, and Tunisia has increased its production by about 2 million tonnes. In 1999, two new phosphate mines opened in Canada and Australia. Increased future production may also come from Australia, Jordan, Iraq and South Africa.

In 1999, production in the FSU countries was approximately one-third of 1988 levels. Recovery from the political changes and economic collapse of the FSU has been slow and may continue to be so in the foreseeable future. Production at current or declining levels is expected for Tunisia, Togo, Algeria, India, Israel, Senegal, Syrian Arab Republic, Brazil and Nauru. Production could come from new mines in Peru, Saudi Arabia, Mauritania and Guinea-Bissau.

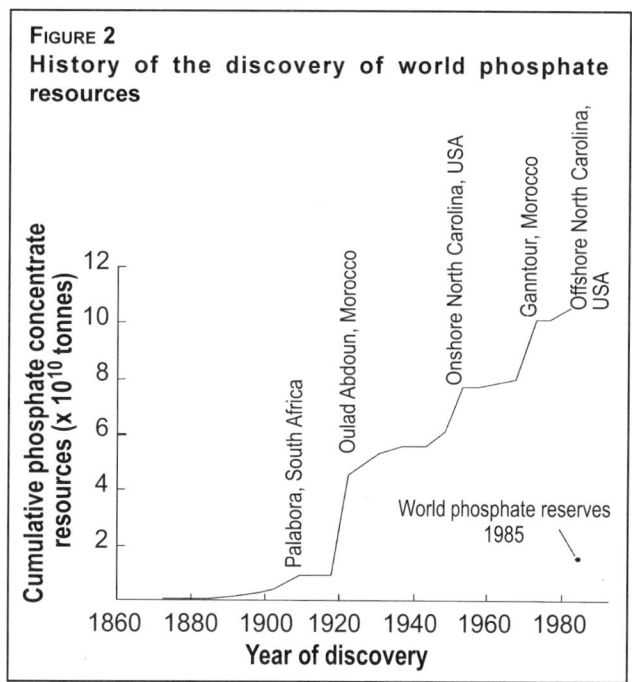

FIGURE 2
History of the discovery of world phosphate resources

Source: Sheldon, 1987.

There is also potential to discover new deposits. In the past 100 years, phosphate has been discovered at a rate (Figure 2) that exceeds the rate of consumption (Sheldon, 1987).

One source of future phosphate production is offshore deposits. Deposits of this type occur along the southeast coast of the United States of America, on the Peru-Chile shelf, off the coast of Namibia, on the Chatham Rise off New Zealand, off the coast of Baja California, Mexico, and off the Congo River delta. None of these offshore deposits is being mined, and they will probably not be mined while ample reserves exist onshore.

Chapter 3
Characterization of phosphate rocks

Characterization studies of phosphate rock (PR) samples should provide data on: (i) the composition of the apatite, other phosphate minerals and gangue minerals; (ii) the relative amounts of mineral species present (estimated); and (iii) the particle size of the various minerals in the rock fabric, etc. It is possible to combine this information with a comprehensive chemical analysis to determine the distribution of chemical species among the mineral components.

Based on such an evaluation, it is possible to estimate the potential for beneficiation, suggest possible beneficiation routes, and make a preliminary assessment of the suitability of PRs for various fertilizer production processes and/or their applicability for direct application.

PHOSPHATE ROCK MINERALOGY

Sedimentary apatites

Most sedimentary deposits contain varieties of carbonate-fluorapatite that are grouped under the collective name francolite (McConnell, 1938). In establishing a series of systematic relationships between francolites, various authors (McClellan and Lehr, 1969; McClellan, 1980; McClellan and Van Kauwenbergh, 1990a) have used X-ray diffraction (XRD), chemical analysis and statistical methods to show that the contents of calcium (Ca), sodium (Na), magnesium (Mg), phosphorus (P), carbon dioxide (CO_2) and fluorine (F) can describe most francolites adequately.

Most importantly, carbonate substitutes for phosphate in a 1:1 ratio (Figure 3), and the maximum amount of substitution is 6–7 percent CO_2 by weight. Both cation and anion substitutions compensate net charge imbalances. The incorporation of CO_2 into the francolite structure is accompanied by increased fluorine contents. In francolites, the unit-cell a dimension (a value) decreases from 9.369 +/- 0.001 Å to approximately 9.320 +/- 0.001 Å with maximum carbonate substitution (Figure 4). The index of refraction also decreases systematically with increasing carbonate substitution.

Sedimentary PRs from insular and cave deposits often contain carbonate apatites that have a lower F content than stoichiometric fluorapatite and may contain significant amounts of hydroxyl in their structures. Although some of these carbonate apatites may meet the francolite definition (significant CO_2 with more than 1 percent F) (McConnell, 1938), they have crystallographic, chemical and other physical properties that differ substantially from those of francolites that contain excess fluorine (McClellan and Van Kauwenbergh, 1990b; Van Kauwenbergh and McClellan, 1990a; Van Kauwenbergh, 1995).

Some carbonate apatites do not fit well within either the excess fluorine francolite series or the hydroxyl-fluor-carbonate apatite series and perhaps belong in an intermediate class. PRs from deposits at Sechura, Peru, and Mejillones, Chile, fit into this category. The carbonate apatites in these rocks have unit-cell a dimensions that fall clearly into the range of the francolite series

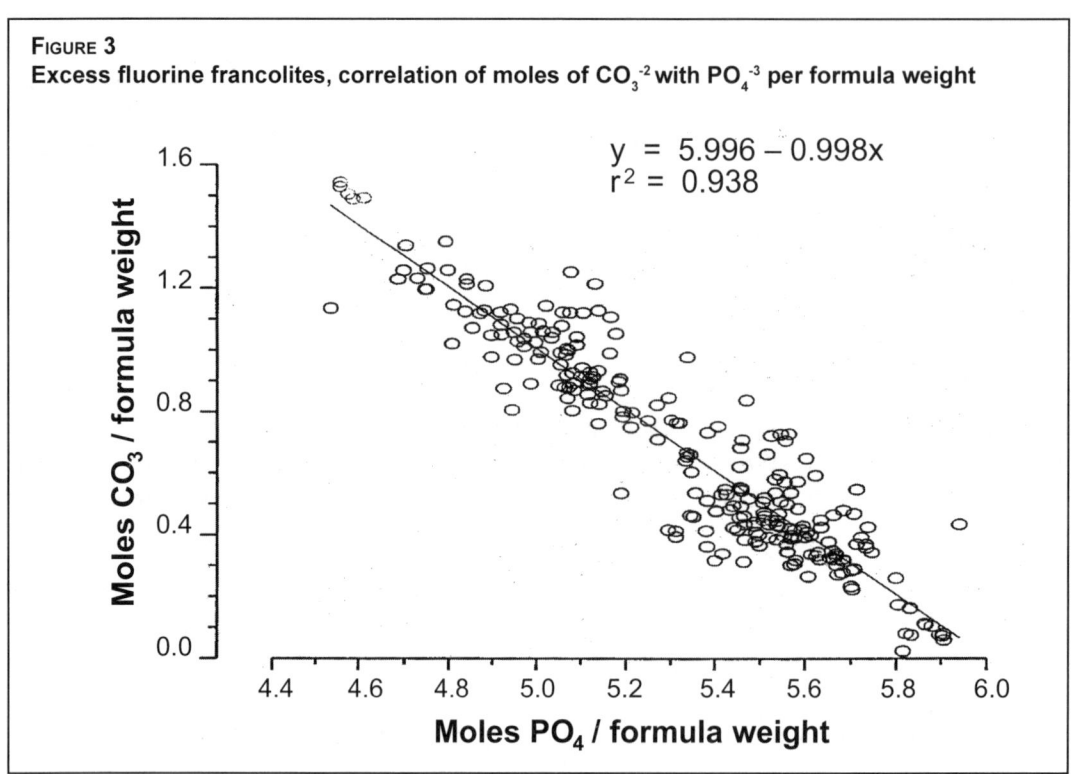

FIGURE 3
Excess fluorine francolites, correlation of moles of CO_3^{-2} with PO_4^{-3} per formula weight

$y = 5.996 - 0.998x$
$r^2 = 0.938$

Source: McClellan and Van Kauwenbergh, 1990a.

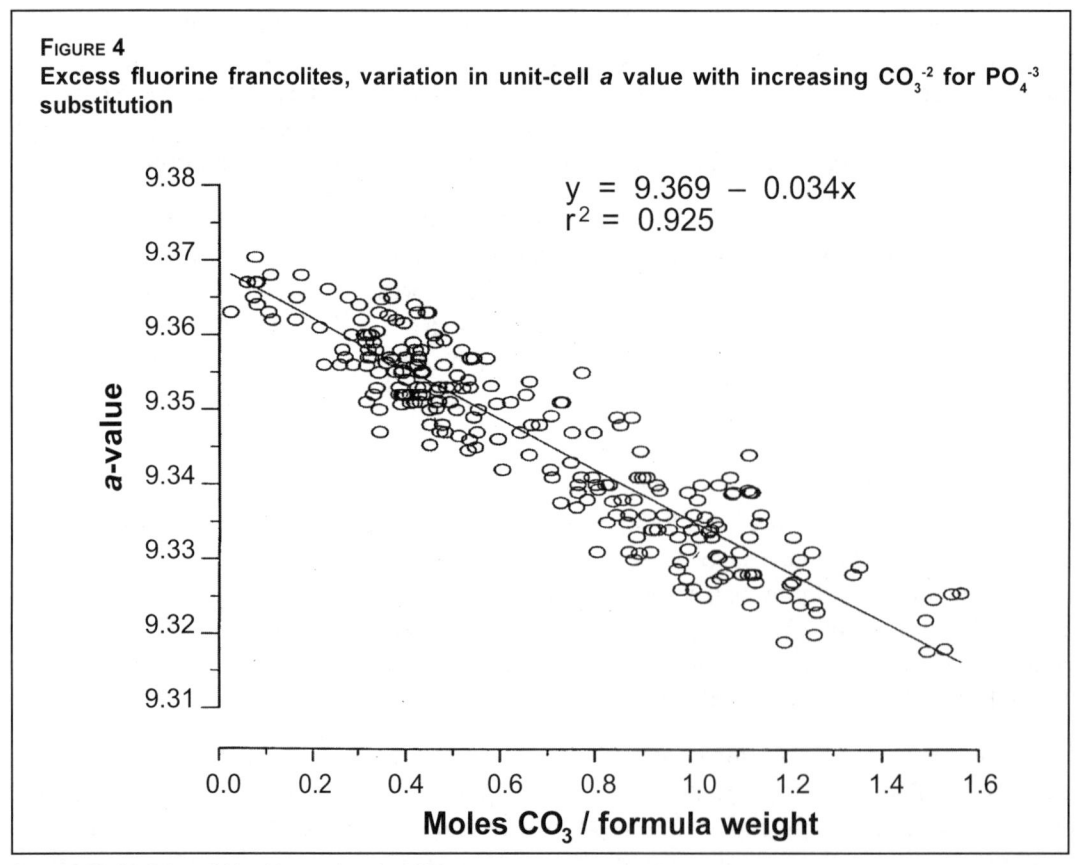

FIGURE 4
Excess fluorine francolites, variation in unit-cell *a* value with increasing CO_3^{-2} for PO_4^{-3} substitution

$y = 9.369 - 0.034x$
$r^2 = 0.925$

Source: McClellan and Van Kauwenbergh, 1990a.

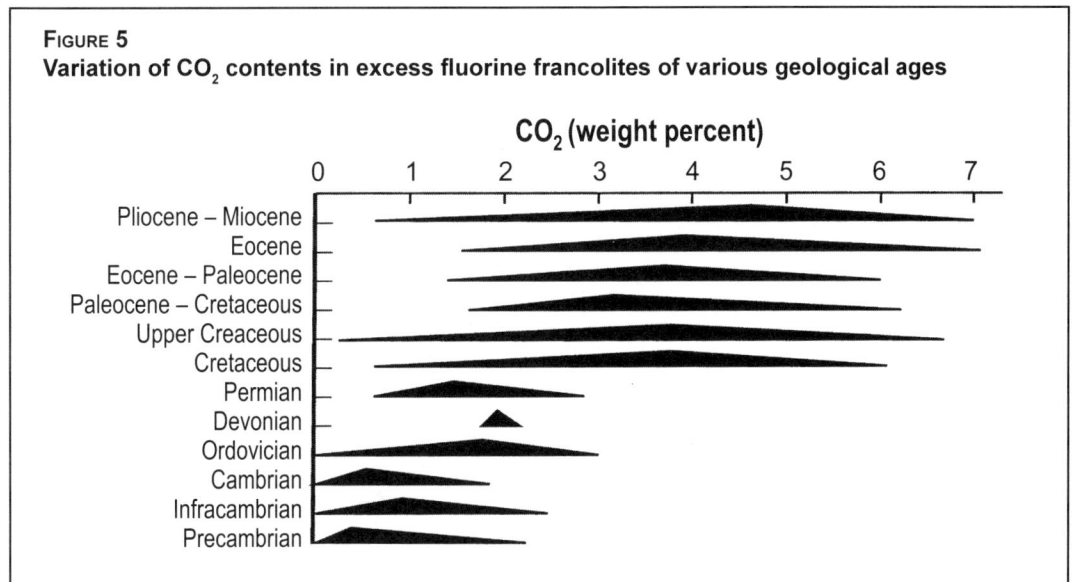

FIGURE 5
Variation of CO_2 contents in excess fluorine francolites of various geological ages

Notes: 472 samples, 165 deposits; peak of triangle represents average CO_2 content.
Source: McClellan and Van Kauwenbergh, 1991.

(9.320–9.370 Å). However, samples from these deposits exhibit higher carbonate contents and chemical reactivities than calculations based on excess fluorine francolite models would indicate.

Francolites are metastable with respect to fluorapatite and can be altered systematically through the combined effects of weathering, metamorphism and time (McClellan, 1980). PRs from the same sedimentary deposit may contain apatites with widely differing properties because of geologic conditions and post-depositional alterations (Van Kauwenbergh and McClellan, 1990b; McClellan and Van Kauwenbergh, 1991). Older sedimentary rocks generally contain francolites with a limited amount of carbonate substitution while younger sedimentary PRs may have compositions that span the francolite model (Figure 5).

Igneous apatites

Primary crystalline apatite from igneous sources may be of fluorapatite, hydroxyapatite or chlorapatite varieties, and pure apatites of the varieties will contain slightly more than 42 percent P_2O_5. A continuous series exists between the fluorapatite and hydroxyapatite end members.

Other minerals in PRs

A total P_2O_5 analysis of a potential ore is not a dependable criterion for estimating the apatite content and evaluating a phosphate deposit. Probably the most common secondary phosphate–weathering minerals are members of the crandallite series. Wavellite is also a common aluminium phosphate formed during weathering. The most common non-phosphatic accessory minerals associated with sedimentary PRs are quartz, clays and carbonates (dolomite and calcite). Carbonate-cemented PRs are particularly noteworthy because of their abundance. Quartz can occur as detrital grains or as microcrystalline varieties. Siliceous cement can be particularly difficult to detect in samples using optical microscopy because of its anisotropic nature and appearance similar to microcrystalline francolite.

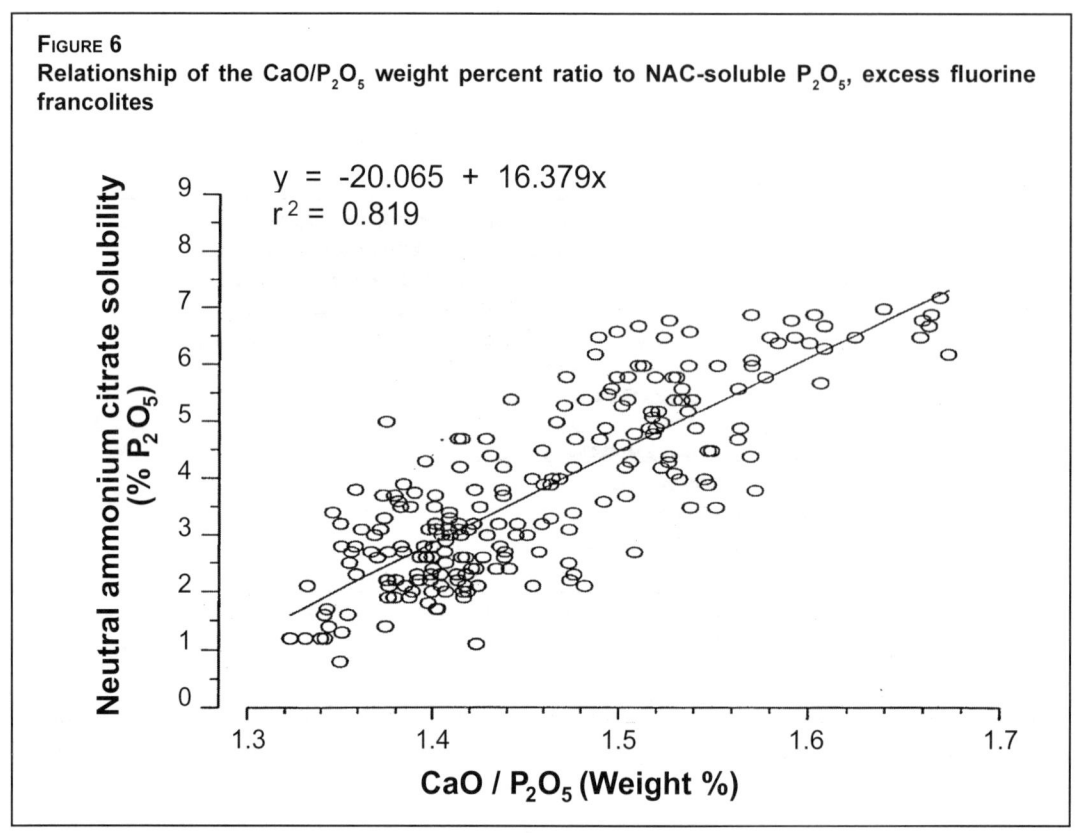

FIGURE 6
Relationship of the CaO/P_2O_5 weight percent ratio to NAC-soluble P_2O_5, excess fluorine francolites

Source: Van Kauwenbergh, 1995.

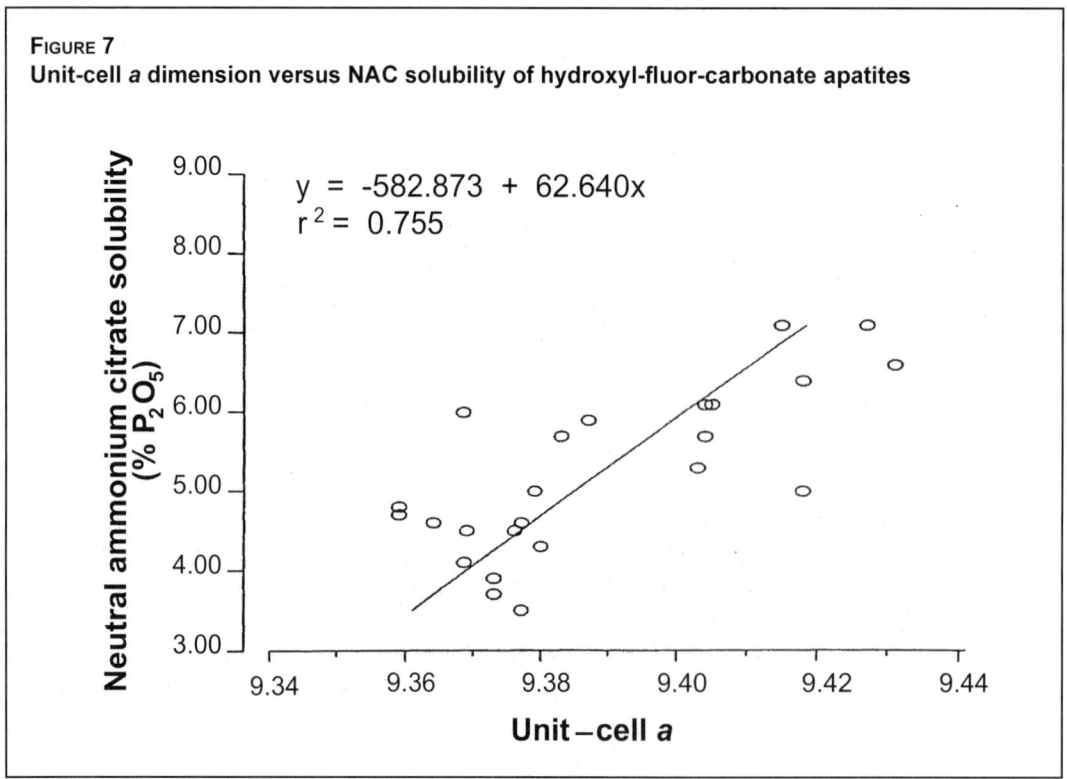

FIGURE 7
Unit-cell a dimension versus NAC solubility of hydroxyl-fluor-carbonate apatites

Source: Van Kauwenbergh, 1995.

Other silicates found in sedimentary PRs can include feldspars and micas (biotite and muscovite). Clay minerals found in sedimentary PRs include illite, kaolinite, smectites and palygorskite/sepiolite. Zeolites, including clinoptilolite and heulandite, are occasionally found in PRs.

Common minerals associated with igneous apatite include nepheline, alkali feldspars, micas, pyroxenes and amphiboles. Calcite, dolomite and magnetite are common minerals associated with carbonate apatite deposits. Weathering may leach and remove carbonates and some silicates leaving resistant minerals such as apatite, magnetite, pyrochlore and zircon in the residuum over the deposit.

PHOSPHATE ROCK SOLUBILITY TESTS

Apatite solubility

The three solutions commonly used to measure the solubility (reactivity) of direct application phosphate rock (DAPR) are neutral ammonium citrate (NAC), 2-percent citric acid (CA), and 2-percent formic acid (FA). The methods used to measure the solubility of PR rock stem from procedures used to analyse conventional water- and citrate-soluble phosphate fertilizers. Unless standard procedural steps are observed strictly, solubility figures obtained by different experimenters on the same rock sources using the same methods may show considerable disagreement (Hammond *et al.*, 1986b).

The NAC solubility of francolites with a maximum known amount of CO_3 substitution ($CaO/P_2O_5 \cong 1.67$) is about 7 percent P_2O_5 (Figure 6). This value decreases with decreasing CO_3 substitution to about 1–2 percent P_2O_5 for sedimentary francolites with very little CO_3 substitution ($CaO/P_2O_5 \cong 1.33$). PRs containing carbonate-apatite with low fluorine contents and OH- substitution may have solubilities in various extraction media as high as the most highly substituted francolites (Figure 7). NAC solubilities of igneous apatites are generally about 1–2 percent P_2O_5 or about the same as sedimentary francolites with little carbonate substitution.

Table 5 shows the data on solubility in NAC, 2-percent CA and 2-percent FA for some selected PRs (in order of decreasing CO_3 substitution). The two PRs that probably show the most consistently high NAC solubilities are North Carolina (the United States of America) and Gafsa (Tunisia). The Central Florida (the United States of America) and Tennessee (the United States of America) samples have less carbonate substitution and lower solubilities. The

TABLE 5
Solubility data on selected phosphate rock samples

Sample	Type	Total P_2O_5	Apatite[a] CO_3 subst.	Solubility[b] (% P_2O_5)		2% citric acid	2% formic acid
				Neutral ammonium citrate			
				1st extraction	2nd extraction		
				(wt. %)			
North Carolina (United States of America)	Sedimentary	29.8	6.4	7.1	6.6	15.8	25.7
Gafsa (Tunisia)	Sedimentary	29.2	5.8	6.6	6.8	11.9	18.6
Central Florida (United States of America)	Sedimentary	32.5	3.2	3.0	3.2	8.5	8.2
Tennessee (United States of America)	Sedimentary	30.0	1.6	2.5	2.7	8.7	6.9
Araxa (Brazil)	Igneous	37.1	0	1.7	1.7	3.5	3.9

a. Based on statistical models developed at the IFDC.
b. All samples ground to -200 Tyler mesh (-75 μm) under similar conditions.

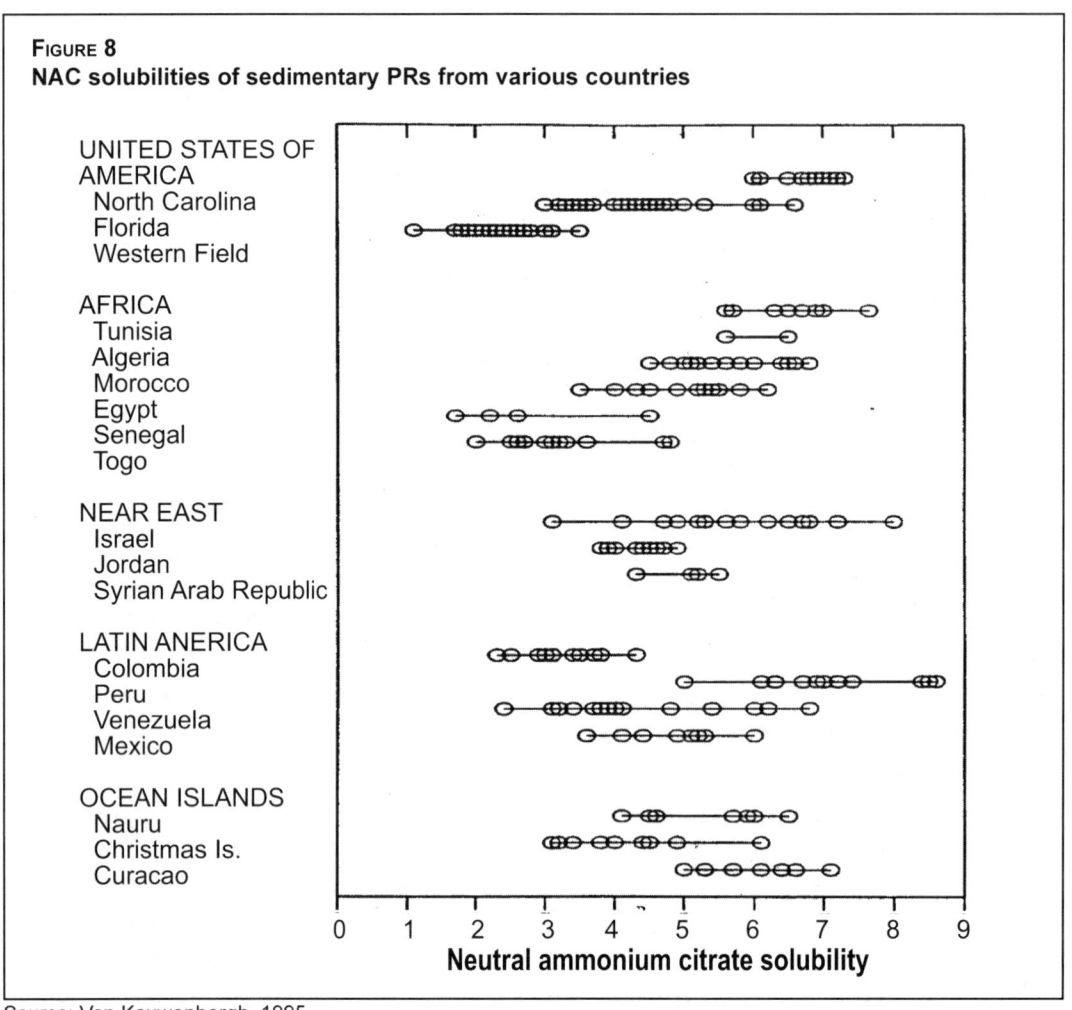

FIGURE 8
NAC solubilities of sedimentary PRs from various countries

Source: Van Kauwenbergh, 1995.

igneous-derived Araxa (Brazil) sample, with essentially no carbonate substitution, has the lowest solubilities in this set of samples.

The measured ranges (maximum and minimum) in NAC solubilities of PRs from various countries may show great variability (Figure 8). Impurities such as calcite, dolomite and gypsum may cause interferences in solubility measurements (Axelrod and Gredinger, 1979). It may be necessary to remove carbonates that interfere with and suppress solubility measurements. It is possible to remove carbonates using Silverman's solution (tri-ammonium citrate solution, pH 8.1) (Silverman et al., 1952). NAC-soluble P_2O_5 values may increase substantially when carbonates are extracted. All the data presented in this chapter were obtained using extracted samples or samples containing no detectable carbonates.

The specific surface area of phosphate particles has a pronounced effect on apparent solubility. Sedimentary phosphate particles containing highly substituted francolites are composed of microcrystalline aggregates. In sedimentary PRs, the geometric outer surface of phosphate particles contributes only a small portion of the specific surface area. The specific surface area, including internal porosity, of sedimentary rocks may be more than 20 times the specific surface area of igneous rocks, which are composed of solid apatite crystals. Studies for the Tennessee Valley Authority (Lehr and McClellan, 1972) indicate that PRs with the highest citrate solubilities often have the highest surface areas.

Grinding provides 'fresh' particle surfaces, increases geometric surface area, and increases solubility measurements. Studies by the International Fertilizer Development Center (IFDC) of two PRs with high NAC solubilities indicated that grinding from plus 200 mesh Tyler (75 µm) to 100-percent minus 200 mesh increased specific surface area from 40 to 50 percent. The NAC P_2O_5 solubilities of these PRs and a third rock increased from about 60 to 120 percent with grinding to minus 200 mesh. However, previous experimental reviews (Rogers *et al.*, 1953; Cooke, 1956; Khasawneh and Doll, 1978) concluded that any increase in PR solubility rarely justified grinding to a size less than 100 mesh for agronomic use.

CLASSIFICATION OF PHOSPHATE ROCKS BASED ON SOLUBILITY

There is no widely accepted simple system for classifying and grading PRs for direct application according to solubility measurements. Diamond (1979) proposed a threefold classification system (low, medium and high reactivity) according to NAC, 2-percent CA and 2-percent FA solubilities (Table 6). The system was based on IFDC data for the relative effectiveness of extraction media and the results of a wide variety of laboratory experiments and field trials.

TABLE 6
Proposed classification of PR for direct application by solubility and expected initial response

Rock potential	Solubility (% P_2O_5)		
	Neutral ammonium citrate	Citric acid	Formic acid
High	> 5.4	> 9.4	> 13.0
Medium	3.2-4.5	6.7-8.4	7.0-10.8
Low	< 2.7	< 6.0	< 5.8

Source: Diamond, 1979.

Hammond and Leon (1983) proposed a system with four solubility rankings (high, medium, low and very low) based on relative agronomic effectiveness (RAE) and NAC-soluble P_2O_5 (Table 7). This study consisted of a relative comparison of PRs from several sedimentary, igneous and metamorphic sources based on a greenhouse experiment using *Panicum maximum* grass grown in an acid soil. These and additional data were analysed further by statistical methods to show that the low and very low classifications were composed of sedimentary francolites with very low carbonate substitution and igneous PRs (Leon *et al.*, 1986). The RAE values depend not only on the inherent PR properties (NAC solubility) but also on soil properties and crop species/varieties (Khasawneh and Doll, 1978).

TABLE 7
Ranking system for some South American PRs by solubility and RAE

Soluble P_2O_5 in NAC (% P_2O_5)	RAE[1] (%)	Solubility ranking
> 5.9	> 90	High
3.4-5.9	90-70	Medium
1.1-3.4	70-30	Low
< 1.1	< 30	Very low

[1] RAE = [(yield of ground PR) - (yield of check)]/[(yield of TSP) - (yield of check)] x 100.
Source: Hammond and Leon, 1983.

The P_2O_5 content and solubility of DAPR constitute what the consumer purchases. However, grade may not be critical in laboratory evaluations of solubility. Lehr and McClellan (1972) showed that NAC solubilized a relatively constant amount of P_2O_5 in successive extractions of the same sample. While the extracted P_2O_5 remained relatively constant, the total P_2O_5 content of the sample decreased. A fairly constant amount of NAC-soluble P_2O_5 was obtained for all mixtures that contained at least 50 percent PR (or about 20 percent P_2O_5). Confusion in interpreting solubility data begins when the extracted P_2O_5 is expressed as a percentage of total P_2O_5. For example, a 30-percent P_2O_5 rock with a 7-percent P_2O_5 NAC extraction value could be expressed as 23 percent of total P_2O_5. A sample containing the same rock (and apatite) could be diluted to 20-percent P_2O_5. At a 7-percent P_2O_5 NAC-extraction value, the NAC-soluble P_2O_5 as a percentage of total P_2O_5 would be 35 percent. This number is higher and might appear better to a prospective buyer. However, the PR would not be any more soluble, and the customer would buy and transport less nutrient and more gangue minerals.

CHARACTERIZATION METHODS

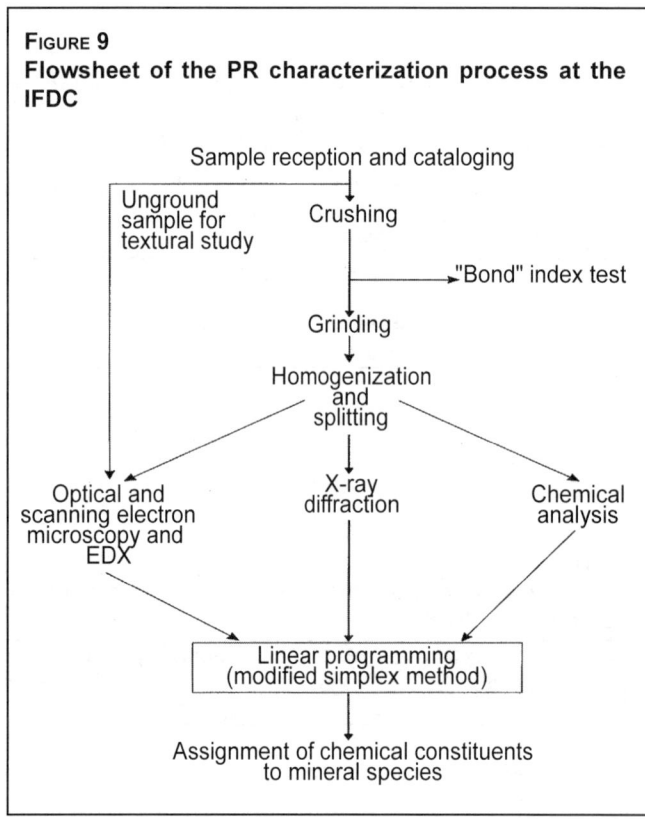

FIGURE 9
Flowsheet of the PR characterization process at the IFDC

The flowsheet in Figure 9 shows an 'ideal' sequence of the steps followed at the IFDC in a detailed characterization of a PR deposit. This scheme is based on experience and can be modified to accommodate specific samples or situations.

Where the sample consists of rock fragments, unground samples are processed to thin sections for optical microscopy and textural studies. The remaining rock fragments are then crushed and ground to be relatively free flowing. These can be homogenized, split and sampled (about minus 0.63 cm). Unconsolidated samples may be screened to determine particle-size distributions, and each size fraction may be analysed chemically.

Microscopic methods include optical and scanning electron microscopy (SEM). Optical microscopy is used to identify mineral species, specify the types of mineral particles, examine the texture of consolidated rocks, and estimate the percentage of various minerals and the liberation and grain size of various components. Samples can be examined as grain mounts or as thin sections.

The texture of PRs can be very important. Consolidated rocks need to be crushed and ground in order to liberate the phosphate particles. Usually, a rock has to be ground to about one-half the grain diameter of the mineral particle of interest or primary gangue mineral in order to achieve the liberation (80 percent mono-mineral particles) needed for beneficiation. Computerized image analysis can be useful in estimating phosphate grain or gangue mineral sizes. While the phosphate grains and gangue mineral grains may occur in some rocks as distinct particles, some rocks contain significant amounts of gangue minerals occluded within phosphate particles.

Cement-particle relationships are also important. If the phosphate grains break cleanly from the cement, it may be possible to make a highly effective separation. If the phosphate grain boundaries and cement are highly intergrown (i.e. locked), separation may be difficult.

When it is not possible to identify minerals by optical microscopy and/or XRD methods, SEM techniques and energy dispersive X-ray analysis (EDX) can be very useful in determining cement/phosphate grain relationships and whether the cement breaks cleanly from the grains. SEM and EDX provide for much higher magnifications and allow close examination of textures of phosphate and gangue mineral surfaces.

XRD is used to identify minerals, qualitatively estimate their concentrations, and determine the unit-cell dimensions of the apatite. Each mineral has a unique XRD pattern composed of a series of peaks. Computer search-match programs ease the complexities of identifying mineral species in complex mixtures considerably.

Special techniques may be needed for clay mineral identification including particle-size separation and concentration. Samples may then be air-dried, treated with organic vapours, and/or also heat-treated to differentiate the clay minerals.

XRD data from apatite and an internal standard can be processed using least-squares computer techniques to determine the unit-cell a and c dimensions. A common problem encountered in the data analysis is the presence of overlapping and interfering peaks. Iron and aluminium phosphates and carbonates (calcite and dolomite) have peaks that overlap with apatite peaks. These peaks may be removed during the course of data processing. Chemical extraction techniques may also be used to remove carbonates prior to XRD analysis.

Chemical analyses of most PRs usually include CaO, P_2O_5, F^-, Cl^-, SiO_2, Al_2O_3, Fe_2O_3, Na_2O, K_2O, MnO, MgO, CO_2, S, organic C, free water, and loss on ignition (LOI). Commercially available standards are used to calibrate analytical methods.

The mineral identifications, determined by XRD, SEM, EDX and optical microscopy methods, and the chemical analyses are equated using several computer methods developed at the IFDC. These methods use a chemical/mineral mass-balance calculation that results in an approximate modal analysis (Figure 10).

The characterization of PR sources, mainly the chemical composition, reactivity and particle size, is the first and essential step in evaluating their suitability for direct application, in particular for comparison purposes in agronomic trials (Chien and Hammond, 1978; Truong *et al.*, 1978; Hammond *et al.*, 1986b). In the framework of an FAO/IAEA networked research project, the agronomic effectiveness of PR sources from several deposits worldwide has been evaluated under different soils, climate and crops conditions (IAEA, 2002). In addition to the individual characterization studies performed by the participating investigators, a complete standard characterization including mineralogical and crystallographic, physical, chemical composition and reactivity analyses was made using 28 PR samples from 15 countries in specialized laboratories. This standard characterization was necessary in order to obtain direct and comparable information on the suitability of the studied PRs for direct application and to better interpret the results from the agronomic evaluation, including the creation of a database for model testing and the development of a decision-support system. Truong and Zapata (2002) have synthesized the information on the methods utilized and results obtained. They have also provided some guidelines for PR characterization for direct application.

FIGURE 10
Diagram showing the approximate modal analysis of phosphorite sample No. 1

Major components	Total	Calcite	Dolomite	Francolite	Pyrite	Quartz	Unassigned
CaO	48.30	0.94	0.25	47.10			
P_2O_5	30.42			30.42			0.00
F	3.78			3.78			0.00
Cl	0.00						0.00
SiO_2	1.79					1.79	0.00
Al_2O_3	0.38						0.38
Fe_2O_3	0.53				0.53		0.00
Na_2O	0.90			0.82			0.08
K_2O	0.10						0.10
MnO	0.00						0.00
MgO	0.47		0.06	0.41			
CO_2	5.83	0.74	0.26	4.28			0.55
S	1.10				0.43		0.67
Total	93.61	1.69	0.57	86.71	0.96	1.79	1.79
-O=F	1.59			1.71			
-O=Cl	0.00						
Total	92.01	1.69	0.57	85.00	0.96	1.79	
Volatiles							
LOI	11.38						
H_2O	1.31						
Organic C	1.38						

Francolite: A = 9.326999
Theoretical Index of Refraction (N25) = 1.604
Theoretical Absolute Citrate Solubility = 17.7%
Predicted AOAC Citrate-Soluble P_2O_5 = 6.3%
CaO/P_2O_5 Ratio, Francolite = 1.55
CaO/P_2O_5 Ratio, Whole Rock = 1.59
CaO/P_2O_5 Ratio, (-Calcite) = 1.56
R_2O_3/P_2O_5 Ratio = 0.03
(R_2O_3 +MgO)/P_2O_5 Ratio = 0.05

Chapter 4
Evaluation of phosphate rocks for direct application

Chapter 3 described the characterization of phosphate rock (PR) materials for direct application. Depending on their origin and geological history, PRs show great variability in their inherent properties, especially in their grade, beneficiation requirements and apatite reactivity. This information provides the first opportunity to assess the suitability of PR materials for direct application. In the case of their utilization in agriculture, PRs applied to soils undergo a series of chemical and biological transformations that govern their dissolution and the availability of the dissolved phosphorus (P) to plants.

This chapter reviews the methodologies for evaluating PRs for direct application in agriculture. They include solubility tests using conventional reagents, incubation studies in soils without plants, pot experiments using a test plant in controlled conditions, and field experiments integrating environmental factors, cropping systems and management practices, as well as their interactions. This chapter illustrates these approaches using selected examples from studies with PRs from West Africa (Truong *et al.*, 1978). The information has been obtained using Taiba PR (Senegal) and Hahotoe PR (Togo), which are exploited on a large scale for export purposes. Other PRs include Arli and Kodjari PRs (Burkina Faso), Tahoua PR (Niger), and Tilemsi PR (Mali), which are mined on a small scale for local use. Gafsa PR (Tunisia) was utilized as a reference because of its high reactivity. Interest in African PRs, in particular those from sub-Saharan Africa, stems from a number of considerations. First, there is the paradoxical situation where Africa ranks first with 28.5 percent of the world's production of PR, yet it has the lowest phosphate consumption with 2.8 percent of the world's consumption (FAO, 1999). Second, although the PR resources of Africa are considerable in terms of both quantity and diversity, they are not exploited greatly (McClellan and Notholt, 1986; Baudet *et al.*, 1986). All types of PR can be found. There are igneous deposits in South Africa, Zambia and Zimbabwe that are coarsely crystalline in nature and quite unreactive and unsuitable for direct application (Khasawneh and Doll, 1978). Guano-type deposits occur in Namibia and Madagascar (Truong *et al.*, 1982). These deposits were formed recently on coral basements and are very soft and practically equivalent to water-soluble phosphate. Finally, there are sedimentary PRs that have been deposited progressively over geological time and are loosely consolidated. They contain microcrystalline particles with large specific surface areas and vary widely in terms of chemical composition and reactivity. These deposits represent 80 percent of the total world reserves. They extend from north to west and central Africa and are among the most relevant for direct application in agriculture.

For recent detailed information, the reader may refer to the FAO/IAEA international networked research project (IAEA, 2002). This project has carried out all types of the studies mentioned above in order to evaluate the agronomic effectiveness of PR sources from several deposits worldwide under a wide range of soils, climate, crops and management conditions.

PHOSPHATE-ROCK SOLUBILITY TESTS

The solubility tests of PRs using chemical extraction methods are empirical. They offer a simple and rapid method for classifying and then selecting PRs according to their potential effectiveness. They cannot be used to evaluate the amount of plant-available P because the actual agronomic effectiveness of the PR in the field depends on a range of factors. Most conventional extraction methods are based on those used at the beginning of the phosphate-industry era (Lenglen, 1935). These include the use of organic acids found in the soil after microbial metabolism and organic matter decomposition, and in root exudates that assist the roots in absorbing phosphates (Amberger, 1978). Water would be the ideal extractant because it is a natural compound (compared with alkaline or acidic electrolytes) and less disturbing of the ionic equilibrium in solid and liquid phases (Van der Paauw, 1971). However, measuring very low P concentrations in the soil solution poses many analytical difficulties. The chemical extractants include reagents such as oxalate, citrate, sodium-ethylenediaminetetraacetic acid, alkaline ammonium citrate and a number of weak acids (including oxalic, lactic, malic, acetic, citric and formic acids), that dissolve and form complexes with the P from the PRs. Various studies have evaluated these methods (Engelstad et al., 1974; Chien and Hammond, 1978; Mackay et al., 1984; Rajan et al., 1992). The most commonly used reagents are neutral ammonium citrate (NAC), 2-percent citric acid (CA) and 2-percent formic acid (FA).

Solubility measurements using conventional chemical techniques

The total P_2O_5 and CaO contents of PRs and their P solubility are obtained using conventional chemical techniques. Table 8 shows such data for African PRs. The total P in PRs ranges from 27 to 36 percent P_2O_5 and the CaO contents are higher than the P_2O_5 values. This is because of the composition of the apatite, which contains 10 Ca atoms for 6 P atoms, and the possible presence of free carbonates. In addition to P, the calcium (Ca) released from PRs can play an important role in the soil ionic equilibrium and in the nutrition of plants grown in acid tropical soils.

The solubility data for the three conventional reagents differ depending on the strength of the extractants. However, they are closely related and rank the PRs in the same relative order: Gafsa, Tilemsi and then the others follow in the same group. Table 8 also shows a good relationship between the solubility in different extractants and the $CO_3:PO_4$ ratio, which indicates the degree of isomorphic substitution of PO_4 by CO_3 in the apatite structure. The greater the degree of substitution, the higher will be the solubility in standard reagents (Chapter 3; Lehr and McClellan, 1972).

TABLE 8
Chemical analysis and solubility in conventional reagents of selected African PRs

Phosphate rocks	Total content % ore		Solubility expressed as % total P_2O_5			Substitution CO_3/PO_4
	P_2O_5	CaO	NAC	Citric acid	Formic acid	
Arli	30.8	47.6	5.4	19.2	38.7	0.098
Kodjari	30.1	44.8	6.1	18.8	37.1	0.093
Tahoua	34.5	44.8	8.3	19.3	34.0	0.112
Taiba	36.5	44.8	5.0	19.8	38.7	0.098
Tilemsi	27.9	43.1	10.4	29.7	47.3	0.210
Hahotoe	35.4	36.4	4.3	19.1	36.7	0.088
Gafsa	30.2	31.9	20.5	37.8	78.6	0.254

Source: Truong et al., 1978.

PR solubility expressions

The expression of P solubility can be an issue. In studies with several PR sources with large variations in total P content, Chien (1993) proposed that the solubility of the PRs is expressed more appropriately as a percentage of the rock rather than as a percentage of the total P (Chapters 3 and 11). This was because unscrupulous practices, such as mixing sand with PR, can inflate the extractability figure (Chapter 11), so leading to incorrect conclusions on the reactivity of PRs.

In general, the solubility expressed as a percentage of total P is preferred for practical reasons. These are that the percentage of the total P that is soluble in a reagent indicates the reactivity of the PR. In addition, the application rate of a P fertilizer is calculated according to the total P content. The expression of PR solubility as a percentage of total P would only be recommended where it represented an inherent property of an individual PR source and/or when comparing several PRs, all PR sources having a similar total P content. However, it is possible to overcome this problem by considering the solubility together with the total P of PRs.

Another expression proposed by Lehr and McClellan (1972) is the absolute solubility index (ASI), which is estimated as follows:

$$ASI = [(\% \text{ soluble P})/(\% \text{ P in apatite})] \times 100$$

The theoretical percentage P content of the apatite can be calculated from the dimensions of the a and c-unit cells. In this way, it is possible to eliminate the variability of the apatite content in the ores and the total P in the apatites (Lehr and McClellan, 1972). The percent-soluble P will depend on the particular test selected. The major disadvantage of the ASI method is that the crystallographic studies are expensive, and they require skilled staff and specialized laboratories with sophisticated equipment for X-ray analysis. Therefore, they should be reserved for basic studies to characterize main phosphate deposits.

Effect of particle size on solubility measurements

As PRs are relatively insoluble minerals, their geometric surface area is an important parameter determining their rate of dissolution. In the case of sedimentary PRs, the geometric surface area is about 5 percent of the total surface area because of the porous structure of PR particles (Lehr and McClellan, 1972). In igneous PR with coarsely crystallised structures possessing no internal surfaces, the geometric and total surface areas are similar.

The PR particle size is also important: the finer the particle size, the greater the geometric surface area and degree of contact between the soil and PR particles and, thus, the greater the PR dissolution rate. The cutoff point for PR particle size appears to be -100 mesh (149 μm), as the cost of finer grinding would be prohibitive compared with any gain in effectiveness. Rajan *et al.* (1992) reported that the increase in chemically extractable P with grinding of PRs related positively to the reactivity of the PRs. They found that the greater the reactivity of the PR, the greater the increase in extractable P on grinding. The data presented in Table 9 show a positive relationship between the fineness of

TABLE 9
Citrate soluble P of North Carolina PR as affected by particle size

Particle size		Total P		Citrate-soluble P
Mesh	μm	%	% rock	% total P
-35	500	13.2	2.7	20.5
-65	230	12.9	2.8	21.7
-100	149	12.9	3.3	25.5
-200	74	13.2	3.8	28.7

Source: Chien and Friesen, 1992.

particle size of North Carolina PR and the increase in citrate-soluble P of the PR (Chien and Friesen, 1992). The message is that measurements of PR solubility should be done on samples of the same particle-size distribution resulting from the same grinding process.

Effect of associated minerals on PR solubility measurements

PRs generally contain endogangue and exogangue materials, such as calcite, dolomite, gypsum, as well as quartz, iron and aluminium oxides and clays (Lehr and McClellan, 1972). Among all of these, free carbonates would have a major influence on PR dissolution because they are more soluble than apatite. They also consume a part of the chemical reagents, especially weak extractants such as NAC, used for the solubility tests. To overcome this problem of preferential reaction in the solubility measurements, Chien and Hammond (1978) discarded the first NAC extract and measured the amount of PR dissolved in the second NAC extract. Mackay *et al.* (1984) and Rajan *et al.* (1992) found the sum of four extracts to be more representative.

Associated minerals have less influence on acidic extractants (Rajan *et al.*, 1992; Chien, 1993). By increasing the strength of the CA from 2 to 5 percent and then to 15 percent, Mackay *et al.* (1984) were able to extract 21, 44 and 59 percent, respectively, of the P in the PR. The influence of free carbonates decreases with the strength of the extractants. Of the three methods that are most commonly used, the 2-percent FA extractant would be preferred for a single chemical extraction procedure.

Kinetics of long-term PR dissolution

Common standard solubility tests are only qualitative in nature. This is because they use procedures that limit the dissolution of PR either by the short reaction time or by the small PR sample/extractant solution ratio. Therefore, these conventional tests focus on the estimation of the short-term efficiency of PRs.

Successive or sequential extraction procedures may improve the predictive capacity of the tests considerably by removing a greater proportion of total P without achieving complete PR dissolution.

Truong and Fayard (1995) have proposed an additional simple procedure for measuring the kinetics of PR dissolution over time. The PR sample (1 g) is placed in a funnel or a tapered tube, blocked with glasswool, and a 2-percent FA solution is dripped continuously over the PR sample at a rate of about 1 drop/s. A preliminary study showed that after 4 h and with 300 ml of solution, most of the P in PRs was dissolved. Figure 11 presents some of the results.

The advantage of this procedure is that it removes all the dissolved P and Ca. Moreover, there is no accumulation on the surface of particles preventing further PR dissolution. The extracted solutions can be analysed at selected time intervals to draw a dissolution curve with time.

In general, the curves shown in Figure 11 present two parts. The first is a zone of fast dissolution for about 50–60 min, which could be related to the short-term efficiency. The curves for different PRs are steeper and close to each other during this part. The second is a zone of slow dissolution with a pronounced change of slope, representing the long-term effect. Some PRs are difficult to dissolve completely because they are cemented by silica or occluded in iron and aluminium oxides (Lehr and McClellan, 1972). In the study shown in Figure 11, the Tilemsi PR confirms its high reactivity in the short and long term. The above procedure is more

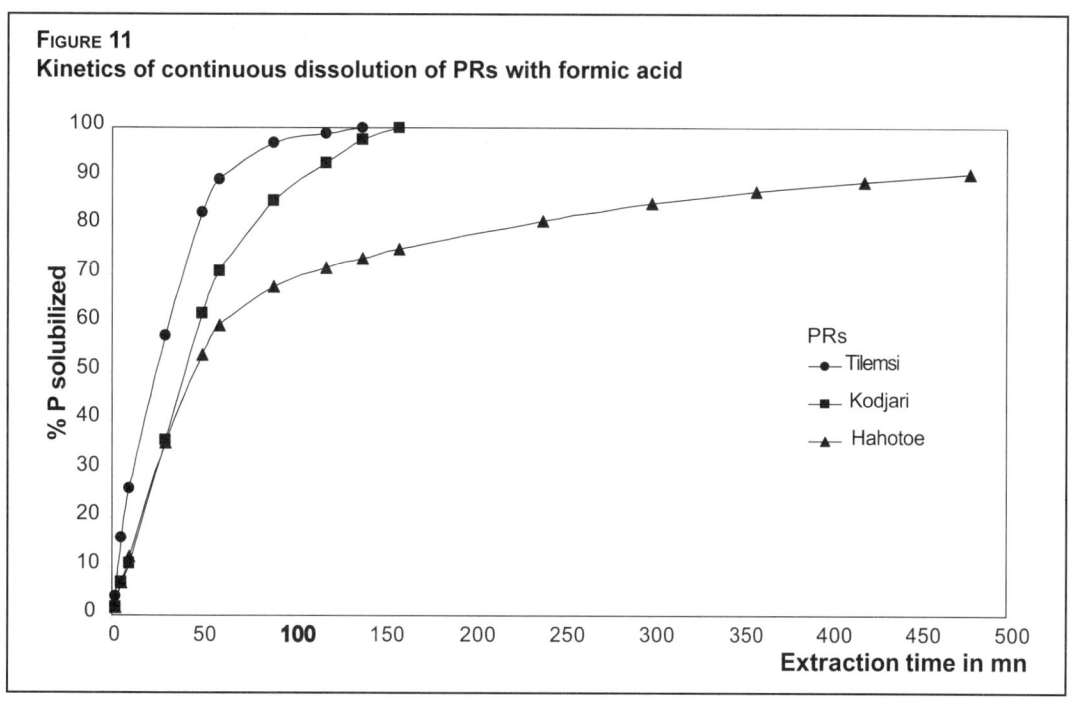

FIGURE 11
Kinetics of continuous dissolution of PRs with formic acid

time consuming than a single extraction. However, it is not more demanding than sequential or successive extractions and it provides more information on PR dissolution over time.

PR reactivity scales and crop yield response

PR reactivity scales serve not only to compare several PR sources but also to predict their potential agronomic effectiveness. Where a reactive PR is applied to a soil, it dissolves under ideal conditions. The available P will result in a good crop response if P supply is a limiting factor. However, the relationship is not direct because many factors and their interactions determine the ultimate agronomic effectiveness of PRs (Mackay *et al.*, 1984; Rajan *et al.*, 1996). Nevertheless, good correlations have been found for a specific set of conditions. For example, the NAC solubility of PRs correlated well with grain yield of flooded rice in Thailand (Engelstad *et al.*, 1974), and dry-matter yield of Guinea grass on an Oxisol from Colombia (Chien and Van Kauwenbergh, 1992). However, several studies have found FA to be the best indicator of crop response (Chien and Hammond, 1978; Mackay *et al.*, 1984; Rajan et *al.*, 1992).

The study by Chien and Hammond (1978) illustrates the value of the FA test. They measured the solubility indices of seven PRs by various laboratory methods and the crop yield response to these PRs on two Colombian soils (Oxisol and Andosol) in greenhouse and field experiments. They found that 2-percent FA gave the highest and most significant correlation coefficient with crop yield response, followed by 2-percent CA and NAC (Table 10). The effectiveness of PRs varied widely with the P-application rate and the duration of the experiment. At low P rates, reactive PRs released P quickly and maintained their dominant position, as less reactive PRs released a small quantity of P slowly. At high rates of P application, reactive PRs released a higher proportion of P at the initial stage and the non-utilized portion was converted into less-available forms in the soil. In contrast, less reactive PRs released a greater P quantity with time, thereby providing an adequate supply to the crop. This improved their effectiveness with time.

TABLE 10
Correlation coefficients between reactivity scales of seven PRs and crop yield

	Neutral ammonium citrate	2% citric acid	2% formic acid
Greenhouse experiment			
Guinea grass			
P added (ppm)			
50	0.65	0.78*	0.76*
100	0.78*	0.86*	0.89*
200	0.89**	0.93**	0.95**
400	0.88**	0.92**	0.96**
Field experiment			
Common beans			
Rate of application, (kg P/ha)			
22	0.75*	0.79*	0.89**
44	0.65	0.71	0.81*
88	0.59	0.65	0.76*
176	0.63	0.74	0.85*

* Significant at 5% level.
** Significant at 1% level.
Source: Chien and Hammond, 1978.

Standardization and index of reactivity

PRs can serve as raw materials for industrial processing in the manufacture of P fertilizers and for direct application in agriculture. The quality criteria for these two uses may be different as the modern fertilizer industry requires strict quality standards for the raw materials that it uses in specific manufacturing processes. Thus, not only the grade (P_2O_5 content) of the PR is important but some other threshold limits should also be considered. For example, the CaO:P_2O_5 ratio is of major importance because of its significance for acid consumption when dissolving the PR. Where the ratio exceeds 1.6, then the wet process is uneconomical. The impurities of aluminium (Al) and iron (Fe) are particularly troublesome in the wet process. The ratio ($Al_2O_3 + Fe_2O_3$):$P_2O_5 > 0.10$ is considered critical. Moreover, a ratio of MgO:$P_2O_5 > 0.022$ is undesirable owing to a detrimental effect on phosphoric-acid production.

The quality factors for direct application are different. Indeed, PR sources suitable for direct application are considered to be 'problem ores' because of their low grade and the presence of accessory minerals and impurities (Hammond *et al.*, 1986b). The presence of carbonates (Ca and magnesium (Mg)) as accessory minerals could be useful for plant nutrition and soil amendment. The Al or Fe contents are usually of no major consequence. Besides suitability for beneficiation, the most important factors in the assessment for direct application are grade (P_2O_5 content) and the reactivity of the apatite (solubility).

TABLE 11
Spatial variability of PR samples within the same deposit

Phosphate deposits	Total P_2O_5, % rock	2% FA solubility, % total P_2O_5
Monte Fresco (Venezuela)		
Intact layer	27.4	10.7
Weathered layer	33.2	16.7
Navay (Venezuela)		
Surface layer	22.6	40.0
Deep layer	19.5	64.4
Matam (Senegal)		
East Block	18.3	20.1
West Block	34.3	71.0

Sources: Truong and Cisse, 1985; Truong and Fayard, 1988.

In some geological deposits, the PR minerals vary within the same deposit. Table 11 presents examples from several deposits. Some layers are more reactive than others. An adequate exploitation of the deposits should consider their quality in relation to whether the PR is to be used for industrial processing or direct application.

The reactivity-index guidelines vary for different countries. The standards of the European Community for the direct application of PRs are strict considering that in Europe most soils are not acidic, annual rainfall is moderate and the cropping is seasonal. To ensure appropriate PR agronomic effectiveness, three kinds of PRs can be sold as fertilizers. These are:

- fine PR: 25 percent total P_2O_5, 55 percent soluble in 2-percent FA, 90 percent < 63 µm, 99 percent < 125 µm;
- semi-fine PR: 25 percent total P_2O_5, 45 percent soluble in 2-percent FA, 90 percent < 160 µm;
- hard PR: 25 percent total P_2O_5, 10 percent soluble in 2-percent FA, 90 percent < 160 µm.

Regarding the United States of America, Chapter 3 presents a classification proposed by Diamond (1979) of PRs for direct application based upon their solubility and expected initial crop yield responses.

In Australia and New Zealand, most PRs are used in permanent pastures and the index of reactivity is set at 30-percent soluble in 2-percent CA (Sale et al., 1997a).

These examples suggest that the reactivity index should be adapted to the local conditions. For example, in the tropics of Latin America and Africa with very acidic soils and heavy rainfall or in Southeast Asia with plantation-state crops, the threshold values of the index could be set lower than those above.

Measurement of exchangeable P from PR by radioisotopic techniques

Isotopic techniques utilizing the beta-emitting radioisotopes ^{32}P (half-life = 14.3 d) or ^{33}P (half-life = 25.3 d) provide a further way of measuring the exchangeable P released from PRs while keeping the ionic equilibrium between the liquid and solid phases unchanged. Although these techniques provide precise, quantitative information, their utilization requires trained staff with adequate skills and expertise, and functional laboratory facilities that comply with radiation-protection and safety regulations (Zapata and Axmann, 1995).

Applications of radioisotopic techniques include the short-term isotopic-exchange kinetics technique to measure soil P dynamics in the laboratory, in particular exchangeable P or E values at selected times (Fardeau, 1981). There are also indirect measures utilizing plants such as labile P or L values (Larsen, 1952) or from isotope dilution, the P derived from PR that is available to the plant (Zapata and Axmann, 1995).

In the African PR study, the exchangeable P values (E values) were determined first. This involved injecting a quantity of carrier-free ^{32}P phosphate ions (R) into a phosphate-water suspension (ratio: 1 g:100 ml). At selected times, 10 ml of the suspension was withdrawn and filtered using millipore filters, and then the concentration of phosphorus (Cp) and the radioactivity (r) of the solution were measured. The isotopically exchangeable P (E value) was calculated using the general isotopic dilution formula: E = Cp x R/r. Table 12 presents the results.

TABLE 12
PR kinetics of isotopic exchange and solubility in water

Phosphate rocks	% ^{32}P remaining in solution			E value after 4 h	Solubility in water
	1 min	10 min	100 min	% total P	% total P
Hahotoe	28.27	15.27	8.98	0.28	0.057
Kodjari	23.60	13.52	7.92	0.17	0.032
Tilemsi	0.58	0.21	0.08	3.87	0.007
Gafsa	0.23	0.10	0.07	3.61	0.006

Source: Truong et al., 1978.

The isotopic-exchange technique can discriminate between PRs in a very short time. After 1 min, two groups can be separated clearly: more than 99 percent of added ^{32}P exchanged with P from Gafsa and Tilemsi PR, whereas only 75 percent exchanged with P from the Kodjari and Hahotoe PRs. Although isotopic exchange is usually a rapid process, the reactions will continue towards equilibrium. After 4 h, the difference between the two groups was confirmed from their E values. The exchangeable P is an indicator of their respective reactivities. However, the differentiation of PRs within the same group (e.g. Gafsa and Tilemsi) requires a longer time of contact of about 24 h (Fardeau, 1993) or several weeks.

The solubility of PR-phosphorus in water was very low for all the phosphates, and the values were not useful in selecting PRs (Table 12).

REACTIONS BETWEEN PHOSPHATE ROCKS AND SOIL

Soil incubation

Incubating soils amended with PRs provides the opportunity to measure the PR dissolution in selected soils with different properties. In addition, some PRs have significant quantities of free carbonates and other minerals and have the potential to modify the characteristics of the soils when the PRs dissolve. Closed-incubation studies to determine PR dissolution rates in soils have limitations because the reaction products are not removed and, therefore, the results could be experimental 'artefacts'. Open-incubation studies, where the soil treated with PR is placed in an open container and water is added at a rate simulating local rainfall conditions, are preferable. The leachate is then collected and analysed for P, Ca and other elements. At selected intervals, soil samples are analysed for dissolved PR.

Closed-incubation studies by Mackay and Syers (1986) showed that the dissolution reaction reached equilibrium at about 50 d. Presumably, at this point, in the absence of any loss mechanisms such as plant uptake or leaching, the P concentration in the soil solution in this study had increased to the point where any further PR dissolution was stopped.

TABLE 13
Effects of incubating PRs on the characteristics of an Oxisol from Njole, Gabon

Treatments	Available P (Olsen-Dabin) mg/kg	pH	Exchang. Ca meq/100g	Exchang. Al meq/100g
Control	31	4.0	0.9	0.12
TSP	110	4.3	1.2	0.07
Hahotoe PR	46	4.1	1.6	0.09
Gafsa PR	76	4.5	2.1	0.01

Source: Jadin and Truong, 1987.

In another incubation study, Jadin and Truong (1987) compared Gafsa, a high-reactivity PR, and Hahotoe, a low-reactivity PR, with a water-soluble fertilizer, triple superphosphate (TSP), including a control without phosphate. The soil used was an Oxisol from Njole, Gabon, with a pH in water of 4.3 and a P sorption capacity of 234 ppm. The P application rate was 100 mg/kg. The available P was measured by the Olsen test as modified by Dabin (1967) to estimate the P dissolved from PRs. The data in Table 13 reflect the net result of opposite reactions occurring upon PR dissolution. These include the release of dissolved P into the soil solution, the sorption of dissolved P by the soil colloids and its conversion to forms non-available to plants. The dissolution rates based on the available P (Olsen-Dabin method) data were 84 percent for TSP, 58 percent for Gafsa PR and 35 percent for Hahotoe PR. The difference between PR sources was consistent with previous evaluation techniques, in particular for the PR reactivity.

PR application prior to planting may be an advantage in soils of low P buffering capacity but not necessarily in soils of higher P buffering capacity (Chien *et al.*, 1990b). The availability of the adsorbed P was measured as a function of the adsorption capacity in order to check the validity of this assumption. In the Njole study, 63 percent of the adsorbed P remained exchangeable during the desorption phase using the ^{32}P isotopic-dilution technique (Jadin and Truong, 1987). Therefore, it is necessary to evaluate the overall P sorption capacity of the soil when considering prior PR applications to the soil.

Liming effect of PRs

Theoretical calculations can estimate the calcium carbonate equivalent (CCE) of PR as the sum of gangue mineral and the carbonate in the apatite. This can also be measured by the Association of Official Analytical Chemists method where 1 g of PR is added to 50 ml of 0.5 NHCl and the remaining acidity is measured by back-titration (Sikora, 2002). Normally, the CCE represents about 50 percent of the PR. These estimates (potential liming effect) assume full PR dissolution.

A greenhouse study assessed the potential Ca value of some PRs from South America and West Africa applied to an Ultisol soil. It found that medium and high reactivity PRs can supply Ca to plants growing in acid soils with low exchangeable Ca (Hellums *et al.*, 1989). In another experiment with an Oxisol from Gabon, the application of Gafsa PR increased soil pH by 0.5 of a unit (Table 13). Higher PR rates can yield larger increases in pH. The effects of the Ca release from PRs were also significant owing to the Ca supply for plant nutrition, its contribution to the base saturation and the decrease in Al toxicity. This result is reflected not only by a decrease in the extractable Al but also by an increase in the level of exchangeable Ca (Kamprath, 1970).

Although the potential increases in soil pH resulting from PR dissolution are small (Sinclair *et al.*, 1993b), they can have a significant effect on Al saturation levels in tropical soils. Work on Ultisols and Oxisols in Puerto Rico showed that Al saturation decreased from 60 percent at pH 4.2 to 35 percent at pH 4.5 and 20 percent at pH 4.8 (Pearson, 1975). These effects can improve soil chemical properties significantly, especially in very acid, degraded and problem soils such as acid-sulphate soils in Southeast Asia (Truong and Montange, 1998).

In conclusion, the liming effect of PR exists but it is of small magnitude. Realistic PR application rates (100–200 kg/ha) with an effective neutralizing value of about 50 percent are equivalent to 50–100 kg of lime per hectare. However, in spite of this small amount, medium and highly reactive PRs can have beneficial effects on the chemical properties of highly weathered tropical soils.

GREENHOUSE TESTS

The next step in the evaluation involves pot experiments using a test plant. They have the advantage of being relatively inexpensive and enabling several factors such as soil type, PR material and plant species (which influence the agronomic effectiveness of PRs), to be examined. This approach can generally control other factors that influence plant growth, such as light, moisture, temperature and disease. This is often difficult in the field. However, growth conditions and plant development in the greenhouse are usually very different from field conditions. For example, the volume, depth and stratification of the natural soil horizons in the field influence the development of the root system in different locations within the profile and regulate the storage and movement of soil water and nutrient ions. These are difficult to reproduce in a pot (Rajan *et al.*, 1996). Considering the slow-release characteristics of PRs, a long-term field evaluation requires variations in soil and climate conditions and in the management practices of the cropping systems.

There are a number of practical considerations that are relevant to conducting pot experiments. The size of the pot depends on the test plant and the expected duration of the experiment. Pots containing 2–5 kg of soil are normally used for small-grain cereals and grain legumes to obtain grain yield. The duration of the experiment depends on the objectives of the study. Some studies recommend harvesting the plants at the period of maximum P demand by the plants, such as at the heading stage for cereals and at flowering or podding for legumes. Others focus on the effect of the treatments on the grain yield and, therefore, will grow crops to maturity. To compare a large number of PRs in a wide range of soils, small grasses are recommended such as *Agrostis* sp. in 100–200 g of soil per pot, or ryegrass in 500–1 000 g of soil per pot. Rapid growth allows several cuts at intervals of 2–4 weeks. *Agrostis* is particularly interesting because of its small-sized seeds (100 seeds/pot weigh 12–15 mg and contain 50–60 μg P) and its responsiveness to the soil P status (Truong and Pichot, 1976). The large number of seeds per pot is important for ensuring homogeneity of germination, tillering and biomass production. The P application rate should vary between 25 and 400 mg per kilogram of soil. Low rates (25 or 50 mg/kg) are often insufficient in soils with low and high P sorption capacities and the effects are not detected. High rates (400 mg/kg or more) can induce disturbances in soil ionic equilibrium (Zapata and Axmann, 1995).

The African PR study by Truong *et al.* (1978) compared seven PRs with different reactivities with a soluble fertilizer (TSP) and a control without P. The P sources were applied at the rate of 100 mg of P per kilogram of soil to pots containing 100 g of soil. The three soils used were an Alfisol from Niger (pH water 6.5; Langmuir's P sorption capacity 16 ppm), a Vertisol from Senegal (pH 5.9, P sorption 1 067 ppm) and an Andosol from Madagascar (pH 4.3, P sorption, 3 818 ppm). They were all deficient in available P. Fertilizers and soils were mixed thoroughly with adequate levels of other nutrients in order to ensure that P was the only factor influencing the results. ^{32}P carrier-free was also added to determine the isotopically exchangeable P (L values) according to Larsen (1952). The *Agrostis* plants were cut every month for a total of four cuts and the dry matter and P uptake were determined. The P-fertilized treatments (TSP and PRs) produced significantly greater dry-matter yields and higher P uptake and L values than the control with TSP giving the highest yields (Table 14). The actual data for the PR treatments varied from soil to soil but the relative ranking of PRs remained almost the same, i.e. Gafsa was the most reactive, followed by Tilemsi, and sometimes Tahoua was better than the others.

TABLE 14
Results of a pot experiment with selected African PRs

Treatments	Dry matter (mg/pot) Sum 4 cuts			P uptake (μg P/pot) Sum 4 cuts			L value (ppm P)		
	Andosol	Vertisol	Alfisol	Andosol	Vertisol	Alfisol	Andosol	Vertisol	Alfisol
Control	635d	1 165f	235e	478e	832f	146e	30f	38e	6e
Arli	1 343c	1 505de	320d	1 968d	1 282e	186cd	77e	39e	28d
Kodjari	1 336c	1 430e	332d	1 720d	1 267e	195cd	74e	39e	30d
Tahoua	1 621bc	1 791d	431c	2 496c	1 956d	269c	100c	46d	33c
Taiba	1 389c	1 365e	335d	2 146cd	1 251e	195cd	90d	41e	28d
Tilemsi	1 769b	1 932c	428c	3 352b	2 700c	259c	134b	62c	33c
Hahotoe	1 313c	1 438e	337d	1 919d	1 361e	209cd	91d	45d	30d
Gafsa	1 892ab	2 260b	506b	3 774b	4 575b	318b	139b	94b	42b
TSP	1 974a	2 862a	1 599a	4 078a	6 549a	4 975a	150a	128a	88a

Values followed by the same letter are not statistically different (P = 0.05).
Source: Truong *et al.*, 1978.

P uptake seems a more sensitive parameter than dry matter for discriminating between PRs. In general, the results confirmed that the inherent characteristics of PRs remain the most important factor determining their potential agronomic value.

In conclusion, pot experiments serve mainly to obtain preliminary information on the potential agronomic effectiveness of PR sources. They are particularly useful for integrating the effects and interactions of PRs, soils and plants under controlled conditions. Nevertheless, the results obtained in these experiments require careful interpretation.

Comparison of PRs with standard fertilizers

The agronomic performance of PRs relative to water-soluble P fertilizers is normally expressed in two ways. The first is the substitution value (SV) or horizontal approach. This is the ratio of the P rate of the standard fertilizer to that of the test fertilizer that is required to produce the same plant yield. The second is the relative response (RR) or vertical approach. This is the ratio of the response to the test fertilizer to that of the standard fertilizer when both are applied at the same rate of P. Both approaches have their advantages and their disadvantages.

The SV is useful for making an economic assessment of the PR fertilizer relative to the reference fertilizer (Chien *et al.*, 1990a) at the desired level of productivity that a farmer may wish to achieve. The SV system has a constant value independent of P rate if the reference and test fertilizers have the same maximum yield, which is rarely the case with PRs (Rajan *et al.*, 1991a, Ratkowsky *et al.*, 1997). For this reason, SV values are calculated at the desired yield level (Rajan, 2002) or as a continuous function of yield (Singaram *et al.*, 1995).

As the yield response curves with PRs and water-soluble P invariably do not share a common limiting yield, it is not possible to calculate a single RR value for each source. The recommendation is to make the evaluation over a full range of P application rates. However, this approach implies a large number of pots or experimental units, and it requires complicated supervision and logistics. A compromise involves working at moderate rates, in the region of the linear part of the response curves, where the RR would be simply the ratio of the slopes of the linear portions of the regression lines (Khasawneh and Doll, 1978). With moderate rates of 50–200 mg of P per kilogram of soil, commonly used in greenhouse experiments depending on soil texture, P status and P-fixing capacity, it has been found that the comparison of various P sources was independent of the rate of P application (Morel and Fardeau, 1989).

Another index is called the relative agronomic effectiveness (RAE) of a given P test fertilizer. This is determined by expressing as a percentage the ratio of the response to the test fertilizer (treatment – control) to the response to the standard fertilizer, when both are applied at the same rate:

$$RAE = [(\text{test P fertilizer} - \text{control})/(\text{standard P fertilizer} - \text{control})] \times 100$$

For the RAE values to be meaningful, the difference between standard P fertilizer and the control should be statistically significant. The PR under study is the P test fertilizer while the standard P fertilizer is a water-soluble P fertilizer such as single or triple superphosphate. Based on the RAE equation, different coefficients can be calculated for crop yield or dry-matter production, P uptake, chemical extraction or L values. Table 15 shows the data obtained for the African PR study utilizing TSP as a reference fertilizer. It shows that the relative ranking of PRs is almost the same with different coefficients, but the actual values vary largely with the soils. Considering the RAEs based on P uptake, Gafsa PR was equivalent (98 percent) to TSP in the Andosol but its relative effectiveness was 54 percent in the Vertisol and 42 percent

TABLE 15
Estimated RAE coefficients based on L values and P uptake from the pot experiment

Phosphate rocks	RAE coefficient (%) (L value)			RAE coefficient (%) (P uptake)		
	Alfisol (Tarna)	Vertisol (Richard Toll)	Andosol (Ambohi-mandroso)	Alfisol (Tarna)	Vertisol (Richard Toll)	Andosol (Ambohi-mandroso)
Arli	27	1	39	27	5	42
Kodjari	29	1	37	32	3	35
Tahoua	33	8	58	38	14	52
Taiba	27	3	50	28	5	42
Tilemsi	33	27	87	38	16	69
Hahotoe	29	8	51	31	6	38
Gafsa	44	62	91	42	54	98

Source: Truong et al., 1978.

in the Alfisol. Conversely, the least reactive PR (Arli) was 42 percent equivalent to TSP in the Andosol. Thus, the ranking of PRs for direct application remains relatively constant, but the effectiveness of the PRs should be considered in relation to the soil properties.

Effect of time

PRs are slow-release fertilizers. They require time and water surrounding the particles in order to enable the dissolution products to diffuse away from the PR particles into the soil volume. The greenhouse evaluation of the African PRs was undertaken to observe the changes occurring with time. The RAE coefficients based on L values and P uptake changed considerably between 1 and 4 months for most PRs in an Andosol (Table 16). Being very reactive, the Gafsa and Tilemsi PRs dissolved rapidly and their effectiveness remained unchanged or increased slightly after 4 months, while less-reactive PRs required time to express their potential effectiveness. The improvement in the relative effectiveness of PRs over time has been attributed to the continuation of the PR dissolution process while a low P concentration is maintained in soil solution. The improvement may also result from the depletion of P from the soluble P fertilizer as a result of P uptake by plants and the conversion of soluble P to less-available P forms. In general, less-reactive PRs need to be ground more finely in order to ensure a greater and longer time of contact between PR and the soil.

Another important factor affecting PR dissolution is the level of soil water content. Table 17 shows the influence of soil water content on P uptake resulting from the application of several PRs in an Oxisol. The effects are more pronounced for less reactive PRs.

TABLE 16
Changes with time of the estimated RAE coefficients for PRs applied to a Madagascar Andosol

Phosphate rocks	RAE coefficient (%) (L values)		RAE coefficient (%) (P uptake)	
	At 1 month	At 4 months	At 1 month	At 4 months
Kodjari	12	39	19	35
Hahotoe	11	56	24	38
Taiba	19	57	29	42
Tahoua	34	62	49	52
Tilemsi	67	89	72	69
Gafsa	91	89	120	98

Source: Truong et al., 1978.

Relationship between PR solubility and crop P uptake

In general, increased PR dissolution is expected to increase available P, and this will result in improved P uptake and crop yield. However, the relationship is not direct because of the many factors and their interactions that affect the agronomic effectiveness of PRs (Mackay *et al.*, 1984, Rajan *et al.*, 1996). Normally, the predicted reactivity of PRs, assessed from solubility tests, should be validated in greenhouse and field experiments. Table 18 presents the correlation coefficients calculated utilizing data from Tables 8 (solubility) and 14 (P uptake). The coefficients are high for both the Andosol and Vertisol but low for the Alfisol. In the Alfisol, the soil pH is almost neutral and the chemical properties are not adequate for PR dissolution.

TABLE 17
P uptake from PRs in a Madagascar Oxisol as influenced by soil water content

Phosphate rock	Phosphorus uptake (µg P/pot)	
	25% field capacity	50% field capacity
Arli	5.22	9.44
Kodjari	5.35	7.72
Tahoua	8.04	11.78
Taiba	6.26	9.48
Tilemsi	13.41	18.22
Hahotoe	4.66	9.68
Gafsa	26.05	23.69
TSP	24.62	25.33

TABLE 18
Correlation coefficients between P uptake and solubility tests

	Andosol	Vertisol	Alfisol
Neutral ammonium citrate	0.90	0.98	0.87
2% citric acid	0.90	0.90	0.73
2% formic acid	0.81	0.91	0.69

Source: Truong *et al.*, 1978.

FIELD EVALUATION

The last step of PR evaluation is to conduct field experiments in representative locations of the region or country under study. Field experiments are essential in order to provide a realistic assessment of the performance of PR in practical farming conditions. A countrywide field evaluation is valuable as the agronomic effectiveness of the PR will be affected by inherent PR properties, soil and climate conditions, cropping systems and farmer practices. However, such a programme requires a large, skilled team and a budget that few countries can afford. The National Reactive Phosphate Rock Project from Australia provides a valid example of such a programme (Sale *et al.*, 1997a).

Several field-evaluation programmes including on-station and on-farm experiments involving multilocation trials have assessed the direct and residual effects of PRs. National programmes have been carried out in Chile and Venezuela (Besoain *et al.*, 1999; Casanova, 1992 and 1995; Zapata *et al.*, 1994), New Zealand (Sinclair *et al.*, 1993c), Brazil (Lopes, 1998), Burkina Faso (Lompo *et al.*, 1995; FAO, 2001b), Mali (Bagayoko and Coulibaly, 1995; FAO, 2001b), Togo (Truong, 1986; FAO, 2001b) and Senegal (Truong and Cisse, 1985; FAO, 2001b). For information on an international programme, the reader may refer to the recent FAO/IAEA networked research project (IAEA, 2002).

The following example illustrates the case of Burkina Faso, where the indigenous Kodjari PR had been studied extensively because of its low reactivity and its difficulty in dissolving with mineral acids (Truong and Fayard, 1987; Frederick *et al.*, 1992). Based on geographical location, soil types, climate conditions and cropping systems, the potential areas for evaluating the direct application of this PR were grouped into three zones with the following main characteristics:

	Zone A	Zone B	Zone C
Geographical location	North and east	Central	West and southwest
Rainfall	< 600 mm	600-800 mm	> 800 mm
Soil	Alfisol, sandy	Oxisol, loamy	Ultisol, loamy-clay
Main crops	Millet	Sorghum	Maize
Fertilization	$23N-25P_2O_5-30K_2O$	$34.5N-25P_2O_5-30K_2O$	$46N-25P_2O_5-30K_2O$

Low fertilizer rates were recommended in consideration of economic reasons and climate risks (drought). Research agronomists and extension staff estimated the fertilization rates according to the predicted response to P application at each site. Utilizing background information from past experiments and considering the available budget and technical capacities of the local staff, the limited number of treatments included: (i) a control NK without P; (ii) NPK with Kodjari PR; (iii) NPK with Kodjari PAPR (50 percent acidulated with sulphuric acid); and (iv) NPK with TSP.

A limited number of on-station-type field experiments were set up in each zone. These experiments involved a statistical design, adequate replications and technical supervision in order to obtain accurate information to serve as a benchmark for the zone. In addition, a large number of farmers were selected to participate in the evaluation programme with the same treatments, but without replications. These on-farm experiments were carried out in order to integrate the variability of farmers' practices. Johnstone and Sinclair (1991) estimated that 40 replicates would be required in order to ensure a 90-percent probability of detecting a difference between two fertilizers that differ in P availability by 10 percent.

Table 19 presents the yield data from the two types of experiment. The average RAE values for the partially acidulated phosphate rock (PAPR) were higher than those for the PR in both types of experiments for all crops. The flooded rice cultivated in basins gave higher results, confirming that water is essential for PR efficiency in the dry savannahs of sub-Saharan Africa. A series of additional experiments evaluated the performance of Kodjari PR. Its solubility was 37 percent in FA (Table 8). The RAE values were 29 percent based on L values and 32 percent based on P uptake (Table 16). The average RAE value based on yield data was 48 percent measured in the field for the whole country (Table 19), ranging from 36 percent in the north to 60 percent in the south, depending on rainfall.

TABLE 19
Crop yield and estimated RAEs from field experiments in Burkina Faso

"On station" experiments						
	Millet		Sorghum		Maize	
Treatments	kg/ha	RAE	kg/ha	RAE	kg/ha	RAE
Control	596		916		2 219	
PR	698	68	1 006	39	2 464	35
PAPR	728	88	1 103	81	2 839	88
TSP	745		1 146		2 919	

"On farm" experiments								
	Millet Av. 70 fields		Sorghum Av. 146 fields		Maize Av. 54 fields		Flooded rice Av. 6 fields	
Treatments	kg/ha	RAE	kg/ha	RAE	kg/ha	RAE	kg/ha	RAE
Control	440		671		1 263		2 036	
PR	598	54	911	50	1 976	77	2 348	80
PAPR	642	70	1 004	70	1 959	76	2 455	108
TSP	728		1 143		2 184		2 422	

Sources: Frederick *et al.*, 1992; Lompo *et al.*, 1994.

Chapter 5
Factors affecting the agronomic effectiveness of phosphate rocks, with a case-study analysis

Chapter 4 reviewed several approaches and methods for the evaluation of phosphate rock (PRs) for direct application. It showed that the final step of such evaluation is to determine their agronomic effectiveness in the field because of the need to integrate a range of factors and interactions in practical farming conditions. The information obtained from the field-evaluation programme should be interpreted in order to provide recommendations on the direct application of PRs for the farmers. This entails a correct understanding of the influence of the various factors that affect the dissolution and the agronomic effectiveness of PRs. The main factors are: reactivity of PRs, soil properties, climate conditions (especially rainfall), and types of crops grown.

This chapter examines the influence of these factors on the agronomic effectiveness of PRs. In addition, it presents case studies on the use of PRs in selected countries representing different geographical regions in order to illustrate how these factors and their interactions influence the agronomic performance of PRs. The countries include: Mali and Madagascar (Africa), India (South Asia), Indonesia (Southeast Asia), New Zealand (Oceania), and Venezuela and Brazil (Latin America). Although there is considerable variation in soil types, agroclimate conditions and crops grown within and across regions and countries, the fundamental factors affecting the agronomic effectiveness of PRs remain the same. As several comprehensive reviews on the use of PRs have already been published (Khasawneh and Doll, 1978; Hammond *et al.*, 1986b; Bolan *et al.*, 1990; Sale and Mokwunye, 1993; Rajan *et al.*, 1996; Chien, 2003), this chapter focuses on the analysis of the significance of the main factors mentioned above.

FACTORS AFFECTING THE AGRONOMIC EFFECTIVENESS OF PHOSPHATE ROCKS

Reactivity of PRs

The reactivity of PRs is a measure of the rate of dissolution of PRs under standard laboratory conditions or in a given soil and under given field conditions (Rajan *et al.*, 1996). It excludes the changes in the rate of dissolution caused by varying soil properties and by plant effects. The chemical composition and particle size of PRs determines their reactivity. PRs of sedimentary origin are generally most reactive and, therefore, suitable for direct application.

The chemical properties that influence the reactivity of PRs are phosphate crystal (apatite) structure and the presence of accessory materials, especially calcium carbonate (Chapters 3 and 8). Increasing the substitution of carbonate to phosphate in the crystal structure generally increases the reactivity of PRs. This substitution results in decreased cell *a* dimensions and also in a weakening of the apatite crystal structure (Lehr and McClellan, 1972; Chien 1977). The most reactive PRs are those with a molar $PO_4:CO_3$ ratio of 3.5–5.

Calcium carbonate is the most abundant accessory mineral in PRs. As calcium carbonate is more soluble than the most chemically reactive phosphate mineral (Silverman *et al.*, 1952), its dissolution increases the calcium (Ca) concentration and pH at the phosphate mineral surface. Thus, it is not surprising that accessory calcium carbonate can reduce the rate of PR dissolution in some soils (Anderson *et al.*, 1985; Robinson *et al.*, 1992). However, under field conditions, leaching and plant uptake may remove Ca ions. The magnitude of the removal by leaching may vary according to soil and climate conditions and to the mode of PR application. For surface-applied fertilizer, the calcium carbonate effect can be minimal even if its content is high, as evidenced for Chatham Rise PR from New Zealand, which contained 27 percent free calcium carbonate (Rajan, 1987). On the other hand, for incorporated PR, more than 15 percent of free calcium carbonate may lessen PR effectiveness in a limed alkaline soil (Habib *et al.*, 1999).

As PRs are relatively insoluble materials, their particle size has an important bearing on their rate of dissolution in soil. The finer the particle size, the greater is the degree of contact between PR and soil and, therefore, the higher is the rate of PR dissolution. Moreover, the increase in the number of PR particles per unit weight of PR applied increases the chances of root hairs intercepting PR particles. Thus, application of PR as finely ground materials (usually less than 0.15 mm) enhances both the rate of dissolution of PRs and the uptake of PR-phosphorus in a given soil. On the negative side, because of their dusty nature, the application of finely ground materials is fraught with practical difficulties.

Among the various methods for measuring the reactivity of PRs (Chapters 3 and 4), a rapid test to determine the reactivity of PRs is to extract PRs with dilute chemical solutions, e.g. 2-percent citric acid (CA), 2-percent formic acid (FA) or neutral ammonium citrate (NAC). The phosphorus (P) extracted is usually expressed as a percentage of total P. In general, the greater the chemical extractability of PRs in these solutions, the higher their reactivity and, therefore, their agronomic effectiveness (Chapter 4). The level of chemical extractability needed for PRs to be agronomically effective varies with soil and climate conditions, especially rainfall. In New Zealand and Australia, the recommended extractability level for permanent pasture is 30 percent of total P soluble in 2-percent CA, whereas in the European Community (EC) it is 55 percent soluble in 2-percent FA for crops. The extractable P values should always be considered together with the total P of PRs.

While chemically extractable P is a good indicator of PR reactivity, some PR deposits may give low indices but still be agronomically effective. For example, although Mussoorie PR from India has a CA-extractable P of 8 percent of total P, it has been found to be agronomically as effective as superphosphate in some soils. The enhanced agronomic effectiveness is attributed to the oxidation of iron sulphide to sulphuric acid and to the localized acidulation of the PR. The iron sulphide is present in intimate association with the apatite crystals. In addition, the PR contains organic carbon (1.14 percent) in the mineralogical composition. This is likely to improve its internal porosity and, therefore, the dissolution of PR-phosphorus (PPCL, 1982).

Soil properties

For a given PR to be agronomically effective, the PR should not only dissolve, but the dissolved PR should also be available to plants. The soil properties that favour the dissolution of PR are low pH (less than 5.5), low solution concentration of Ca ions, low P fertility levels and high organic-matter content.

Soil acidity

The dissolution of PR may be expressed by the equation:

$$Ca_{10}(PO_4)_6F_2 + 12H_2O \rightarrow 10Ca^{2+} + 6H_2PO_4^- + 2F^- + 12OH^-$$

(Phosphate rock) (Water)   (Dissociation products)

Although the above reaction is for a fluorapatite PR, it applies to other members of the apatite minerals including reactive PRs (RPRs). As indicated in the above equation, the dissolution of PR results in the release of hydroxyl ions into solution. Neutralization of the hydroxyl ions released by soil acidity enables the PR dissolution process to continue. In the case of PRs where phosphate has been substituted with carbonate ions, hydrogen ions may also be needed to neutralize hydroxyl ions formed on the release of carbonate ions into solution (Chien, 1977b). Each carbonate ion (CO_3^{2-}) joins with two hydrogen ions and forms one water molecule and carbon dioxide gas. Thus, an adequate supply of hydrogen ions is of primary importance for the continued dissolution of PR (Chapter 4).

Indicators of hydrogen ion supply are soil pH and titratable acidity. Soil pH shows the magnitude of hydrogen ion supply at a given time, whereas titratable acidity indicates the supply of hydrogen ions in the longer term. A linear positive relationship has been reported between initial pH and titratable acidity in Australian soils (Kanabo and Gilkes, 1987). As a simple guideline, the use of PRs, depending on their reactivity, is generally recommended in soils with a pH of 5.5 or less. The dissolution of PR diminishes with increasing pH up to 5.5 but the decline is more rapid above this pH level (Bolan and Hedley, 1990; and Chapter 4). When considering a large number of soils, titratable acidity may be a better indicator of PR dissolution (Babare *et al.*, 1997).

Cation exchange capacity, and exchangeable calcium and magnesium

For continual dissolution of PR, it is important that the other major reaction product, the Ca ion, be removed or that its concentration in soil solution be maintained at a lower level than that in the film surrounding the dissolving PR particle. It is possible to achieve these outcomes if there are adequate soil cation exchange sites available to adsorb the Ca ions released from the PR, or if Ca is leached away from the site of PR dissolution. A measure of the cation exchange sites available for Ca adsorption is the difference between the cation exchange capacity of soils and the exchangeable Ca (Bolan *et al.*, 1990; Robinson and Syers, 1991).

Recent studies suggest that high exchangeable magnesium (Mg) in soils may enhance PR dissolution (Perrott, 2003). Theory would suggest that, as Mg is held by soils more strongly than Ca, the presence of Mg on the soil exchange sites can block adsorption of Ca released on dissolution of PR and thereby facilitate its removal from the soil-fertilizer system. This will have the effect of enhanced PR dissolution. In soils with a low pH (less than 5.5), the exchangeable Ca and Mg will invariably be low (low base saturation) and, therefore, there will be low solution concentrations of these ions.

The cation exchange capacity of soils is also related closely to soil texture. Sandy soils usually have a low cation exchange capacity and, therefore, do not provide an adequate sink for Ca released from PR. This would lead to a reduction in PR dissolution and in agronomic effectiveness. The other two scenarios occur in areas of sufficient rainfall. The first is where the released Ca may be removed from near the PR particles, with a positive effect on PR dissolution and on agronomic effectiveness. The second is where excess rainfall may lead to leaching of P below the rooting zone of crops and reduce the agronomic effectiveness of PRs.

However, because of their slow-release nature, PRs are likely to be more beneficial under such circumstances than water-soluble fertilizers (Bolland *et al.*, 1995).

Soil solution P concentration and P retention capacity

As the P concentration in soil solutions is usually very low (0.05–0.5 mg/litre), it has little influence on the dissolution of PR. Nevertheless, there have been reports that the greater the P sorption capacity of soils, which results in depletion of soil solution P, the greater the dissolution of PR (Chien *et al.*, 1980a). It is not P adsorption capacity per se that affects PR dissolution but the number of sites available to adsorb the P released from PR and, therefore, maintain a lower P concentration in solution near PR particles.

When considering a large number of soils, the variation in the rate of PR dissolution in soils can be better explained by taking into account the P sorption capacity of soils in addition to the titratable acidity (Babare *et al.*, 1997). Although increased P sorption capacity may favour PR dissolution, its availability will depend on soil P status and the amount of PR added.

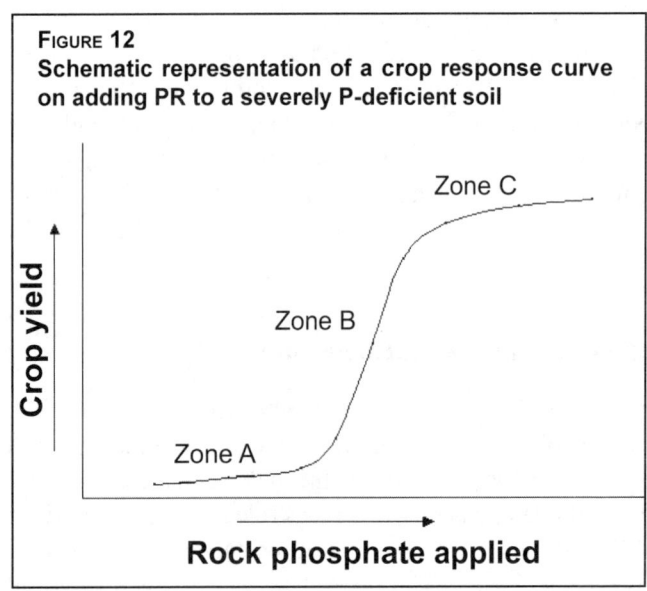

FIGURE 12
Schematic representation of a crop response curve on adding PR to a severely P-deficient soil

When low amounts of PR are added to severely P-deficient soils, the soils adsorb tightly almost all of the dissolved P with very little increase in soil solution P. This results in very little increase in crop production (Zone A in Figure 12). At higher levels of P application, as the solution P increases above the threshold concentration for net P uptake by plants, crop yield rises steeply (Zone B in Figure 12) (Rajan, 1973; Fox *et al.*, 1986). Soils of medium P status are likely to be in the region corresponding to the starting point of Zone B. In this case, the dissolved PR is likely to translate into crop yield. Thus, soils should preferably have a 'medium or above' level of P fertility to bring about immediate benefit from the application of PR at the maintenance rate. The maintenance rate is equivalent to the amount of P removed by crop produce. In such soils, the plant-available P can serve as starter P for crop establishment and early growth, which in turn assists the roots in utilizing PR more effectively. This is similar to the effect of water-soluble P on PR effectiveness (Chapter 9). Zone C in Figure 12 represents the yield plateau attained at high levels of PR application.

Soil organic matter

Another soil property that increases PR dissolution and its availability to plants is soil organic matter (Johnston, 1954b; Chien *et al.*, 1990b). This seems to arise from: (i) the high cation exchange capacity of organic matter; (ii) the formation of Ca-organic-matter complexes; and (iii) organic acids dissolving PR and blocking soil P sorption sites.

The cation exchange capacity of organic matter is greater than that for clay minerals. Depending on their clay content, the cation exchange capacity of mineral soils may range from a few to 60 cmol/kg, whereas that of organic matter may exceed 200 cmol/kg (Helling *et al.*, 1964). The high cation exchange capacity of organic matter means increased Ca retention capacity of soils, which leads to enhanced PR dissolution. Humic and fulvic fractions of organic matter form complexes with Ca (Schnitzer and Skinner, 1969), which can also reduce Ca concentration in solution, so leading to enhanced PR dissolution. The organic matter content of tropical soils is generally less than 2 percent.

When arable crops are harvested, a large proportion of the root residues and in some cases part of the above-ground portions are left behind in the soil. The decomposition of plant residues in soil results in the production of numerous organic acids, such as oxalic, citric and tartaric acids (Chapter 9). These acids can be expected to dissolve PR by supplying the hydrogen ions needed to neutralize the hydroxyl ions produced when PR dissolves and by forming complexes with cations, especially the Ca from PRs. The organic ions and humus can also reduce the P sorption capacity of soils by blocking P sorption sites and by forming complexes with iron and aluminium hydrous oxides, leading to increased P concentration in solution (Manickam, 1993).

Logic would dictate that incorporation of PR during cultivation between crops would benefit the farmers most. Such a practice would allow the decomposing plant residues, and any animal litter that might have been applied, to enhance the release of PR. The early application of PR would also allow time for the reaction of PR with the soil and the release of some P before the next crop is established.

Climate conditions

Rainfall is the most important climate factor that influences PR dissolution and its agronomic effectiveness. Increased soil water brought about by rainfall or irrigation increases PR dissolution (Weil *et al.*, 1994). The process is affected by speedy neutralization of the hydroxyl ions released and removal of Ca and other reaction products from the area adjacent to PR particles. Adequate water supply will encourage plant growth and P uptake by plants, so leading to the increased agronomic effectiveness of PRs. For surface applied PRs, experience in Australia and New Zealand would indicate an annual rainfall requirement of at least 850 mm for PRs to be agronomically similar to water-soluble fertilizers (Hedley and Bolan, 1997; Sale *et al.*, 1997b). However, the rainfall requirement does depend on soil properties.

Chien *et al.* (1980b) reported that temperature has negligible or no influence on the solubility of PRs within a range of 5–35 °C and, therefore, on its agronomic effectiveness.

Crop species

Plant species differ in their P uptake demand and pattern as well as in their ability to absorb soil solution P (Helyar, 1998; Baligar, 2001). Moreover, plant species show differences in their ability to access sparingly forms of P that are unavailable to other plants (Hocking *et al.*, 1997; Hinsinger, 1998; Hocking, 2001). Among these, some plants can dissolve and take up the products from PR dissolution (Hinsinger and Gilkes, 1997). For example, perennial pastures, tree crops and plantation crops require a steady supply of P over an extended time span. Because PRs in soil dissolve gradually and supply P at a steady rate, increasing amounts of PRs are being applied as phosphate fertilizers for the above-mentioned crops (Ling *et al.*, 1990; Pushparajah *et al.*, 1990; Chew *et al.*, 1992). The high agronomic effectiveness of PRs realized with these crops reflects partly the acidic nature of the soils and the high root density. High root density facilitates the intensive exploration of a large soil volume for P because of the presence of a large number of fine roots per unit of soil volume.

Legumes are particularly suited for the use of PRs. They are effective in dissolving PR and in absorbing its dissolution products because of their demand for Ca and the acidifying effect of nitrogen (N) fixation in the soil near the root system (rhizosphere) (Ankomah *et al.*, 1995; Kamh *et al.*, 1999). This effect can be utilized to improve the P nutrition of a companion crop (intercropping) or that of the subsequent crop in a rotation (Horst and Waschkies, 1987; Vanlauwe *et al.*, 2000).

Some plant species (e.g. rapeseed, lupines and pigeon pea) have been studied because of their ability to secrete organic acids that result in an enhanced dissolution of PR (Jones, 1998; Hoffland, 1992; Adams and Pate, 1992; Ae *et al.*, 1990; Montenegro and Zapata, 2002). Recent studies (Chien, 2003) indicate that reactive PRs may have potential applications even in alkaline soils with organic-acid secreting crops such as rapeseed (canola) (Chapter 9).

Crops that possess high Ca uptake capacity are more suited for PR use. In this respect, finger millet is most suited for PR use, followed by pearl millet and maize (Flach *et al.*, 1987).

Management practices

Four important management practices that can influence the agronomic effectiveness of PRs are: the placement of PR material in relation to the plants; the rate of application; the timing of application; and lime application.

PR placement

In order to achieve the maximum agronomic effectiveness from PRs, the material should preferably be broadcast and incorporated uniformly into the surface soil to the required depth. The depth of incorporation for seasonal crops may be of the order of 100–150 mm. Incorporation facilitates greater dissolution of PR by increasing contact between the soil and PR particles. It also enhances plant absorption of P by providing a greater volume of P-enriched soil. In addition, there is a greater likelihood that a root will encounter a dissolving PR particle.

Rate of PR application

The decision on the rate of PR application needs to be based on the soil P status as indicated by soil testing (Chapter 6; Perrott *et al.*, 1993; Perrott and Wise, 2000), and the expected rate of dissolution of PR and its availability to plants (Chapter 9; Perrott *et al.*, 1996; Rajan *et al.*, 1996). The soil testing method to be used would depend on whether the phosphate fertilizer applied previously was a water-soluble form or a PR. Some general guidelines are:

- PR application is likely to be beneficial in soils of medium P status. In such soils, the minimum rate of application should be such that the expected amount of P dissolved from PR is no less than the amount of P removed from the site as farm produce plus the amount of phosphate retained by soils in a form that is not available to crops under near maximum production levels. This is often called the 'maintenance P application' rate. In terms of absolute amounts, soils with high P retention capacities will require greater rates of P fertilizer application than soils of lower P retention. This allows for the P retained by soils in a form that is 'not available' to plants.

- In soils of low P status there are two choices: (i) bring the fertility level to medium P status by applying water-soluble fertilizers and follow this by the application of PR; (ii) incorporate large applications of PR (500–1 000 kg/ha) followed by a regular maintenance application of P.

Timing of PR application

In very acid soils (pH less than 5.5) with a high P retention capacity, the incorporation of PR close to planting time is recommended in order to minimize conversion of dissolved P to plant 'unavailable' forms (Chien *et al.*, 1990b). However, in less acid soils (pH of about 5.5–6) with a low P retention capacity, incorporation of PR 4–8 weeks ahead of planting is preferred. This allows time for some dissolution of PR and its subsequent availability to plants. Laboratory experiments have shown that it may take 4–8 weeks for PRs to reach their maximum solubility (Barnes and Kamprath, 1975).

The use of PR for flooded rice requires special attention because soil pH generally increases upon flooding. For this reason, it is advisable to apply PR to the soil about two weeks before flooding (Hellums, 1991).

Lime application

Incorporating lime has an adverse effect on PR dissolution in soil because it increases the Ca concentration in solution and reduces soil acidity. However, liming may increase the availability to crops of dissolved P by increasing soil pH and reducing aluminium (Al) toxicity. In view of the above effects, where liming is to be done in order to raise soil pH values to 5.5, it can be applied at the same time as the PR application, but preferably not as an admixture with PR. This can eliminate Al toxicity while still encouraging PR dissolution. Where the soil pH is to be raised above pH 5.5, liming should preferably be done about six months after the incorporation of PR so that PR dissolution is not reduced drastically.

CASE STUDIES

Mali

The agriculture sector in Mali involves 80 percent of the population. However, less than 10 percent of the 2.7 million ha of cultivated land receives fertilizers. The country imports 70 000 tonnes of fertilizer per year. This fertilizer is applied mainly on cash crops such as cotton, rice and groundnut. This amount is less than 15 percent of the nutrients exported by the crops. The authorities have long tried to improve the situation, particularly by using local PR.

Tilemsi PR production

Mali has a relatively large phosphate ore deposit of sedimentary origin, located in the Tilemsi Valley about 120 km north of Bourem. Klockner Industrie Anlagen GMBH (1968) undertook a detailed study for its exploitation. The reserve was estimated to be 20 million tonnes with an average P_2O_5 content of 27–28 percent. The PR is easy to extract with a thin overburden. The exploitation started in 1976 as part of a Malian-German technical cooperation project. This involved the installation at Bourem of: an electricity generating unit; grinding, cycloning and bagging machines; and storage facilities. The production capacity was 36 000 tonnes/year.

Actual production was about 1 000 tonnes/year in the 1970s, 3 000 tonnes/year in the 1980s and 10 000 tonnes/year in the 1990s. Production peaked in 1991 at 18 560 tonnes of PR. Since then, production has become inconsistent as a result of political unrest in the mine area.

At an output rate of 10 000 tonnes/year, the estimated production cost was US$20.6/tonne of raw material arriving at Bourem. This increased to US$78.5/tonne for ground and bagged product ex-factory, and then increased to US$157/tonne delivered in Sikasso, the main utilization zone, 1 300 km to the south. The estimated price of triple superphosphate (TSP) delivered in Sikasso was US$273/tonne, which is US$0.60/kg of P_2O_5 compared with US$0.54/kg of P_2O_5 for the Tilemsi PR. This difference is not large (Truong and Fayard, 1993).

A cost analysis shows that the cost of grinding and bagging (US$57.9/tonne) at Bourem is too high. Bourem is located in the desert, so all the components for production including fuel oil, bags, spare parts, maintenance, and the workforce, are very expensive. The transfer of the factory to Segou or Koulikoro 1 100 km to the south could generate important economies. Connection to the national hydroelectricity grid is possible, and many facilities of a big city are available, notably the security needed to ensure regular production. The transport cost is also very high (US$78.5/tonne). The traditional dugouts (capacity 10–20 tonnes) can sail year round on the Niger River. Almost 95 percent of the transported goods flow from south to north, and only 5 percent in the reverse direction. Thus, it would be possible to take advantage of this return freight availability to transport Tilemsi PR at a lower cost.

Tilemsi PR evaluation and use

Tilemsi PR is a medium-reactive rock, with a total P_2O_5 of 29 percent, of which 61 percent is soluble in FA. Therefore, it is suitable for direct application. Since 1977, many studies have been conducted in controlled and natural conditions covering the main pedoclimatic zones and cropping systems in the country. The results have shown that the effectiveness of Tilemsi PR is very dependent on rainfall distribution. The relative agronomic effectiveness (RAE) averages about 80 percent compared with TSP.

Research institutions recommend an annual application of Tilemsi PR of 100–200 kg/ha, or 300–400 kg/ha for a rotation. The PR should be applied on the fallow and incorporated through late ploughing at the end of the rainy season. The Government of Mali has tried to promote the use of Tilemsi PR through national and provincial extension services, and through development companies for cotton, groundnut, rice and sugar-cane crops. The only company to sell significant quantities (about 5 000–10 000 tonnes/year) of Tilemsi PR to farmers has been the Malian Company for the Development of Textiles (CMDT). The CMDT plays an important technical, social and economic role in the region.

TABLE 20
Yield of millet, groundnut, sorghum, cotton and maize with Tilemsi PR and TSP, Mali, 1982–87

Fertilizer treatment	Grain yield kg/ha/year	Productivity kg/P_2O_5/year
Control with N and K applied	676	
Tilemsi PR basal application 120 kg P_2O_5/ha	1 110	3.6
TSP annual application 30 kg P_2O_5/ha	1 302	3.4

In the period 1982–87, the Government of Mali conducted a fertilizer project through the Institut d'Eeconomie Rurale and the International Fertilizer Development Center (IFDC) in the five main pedoecological regions, namely: Mopti (millet), Kayes (groundnut), Segou (millet), Koulikoro (sorghum) and Sikasso (cotton and maize). Henao and

Baanante (1999) made a comprehensive analysis of the results, including an economic evaluation. Table 20 presents a summary of the agronomic results. In medium to long-term experiments, Tilemsi PR is practically equivalent to TSP per unit of P_2O_5.

Madagascar

The fertilizer situation in Madagascar is a cause for serious concern. The total consumption is about 15 000 tonnes/year for 1.7 million ha of cultivated land. The export of nutrients by crops amounts to 205 000 tonnes ($N + P_2O_5 + K_2O$). This means that the replacement rate covers less than 4 percent of the nutrients removed. Therefore, heavy soil mining is in progress.

In Madagascar, PR deposits are limited, with estimated reserves of 600 000 tonnes. These are in the form of guano materials deposited on coral reefs, which are distributed over the island archipelago known as the Barren Islands (Ratsimbazafy, 1975). More than half of these deposits (312 000 tonnes) are located on the island of Andrano. Because these deposits are located on coral reefs, the mining requires special efforts in order to preserve the ecology of the archipelago. This limits mining to 10 000 tonnes/year and requires replacement of the mined PR with the same quantity of soil, which has to be transported from the main island to the coral reefs.

The Barren Island PRs are reactive. For example, the FA solubilities (percentage of total P_2O_5) of the Andrano, Androtra and Morombe PRs all exceed 70 percent. The agronomic effectiveness of these PRs was further confirmed by a pot experiment where the three PRs were applied at the rate of 100 mg of P per kilogram of soil in 100 g of an acid (pH of 4.3) Andosol soil from Madagascar. Three monthly harvests of the test plant *Agrostis* sp. were taken. The availability coefficients for these PRs, defined as: ((P uptake [PR] - P uptake [control])/(P uptake [TSP] - P uptake [control])) x 100, were all in excess of 100 after one harvest, and over the three harvests. These results show that the Barren Islands PRs are very reactive, and that they are equivalent or superior to TSP in terms of agronomic effectiveness (Truong *et al.*, 1982).

Two long-term field trials were conducted on the High Plateau of Madagascar, where the soil pH is 4.5 and the annual rainfall 1 200 mm. These involved growing maize for eight years and paddy rice for nine years. The PR was applied at a rate of 400 kg of P_2O_5 per hectare for maize and 300 kg of P_2O_5 per hectare for rice. There were marked responses in yield above the control treatment over time. These amounted to 25 000 kg of maize over eight years and 5 020 kg rice over nine years (IFDC-CIRAD-ICRAF-NORAGRIC, 1996). Thus, the reactivity of PR and favourable soil and climate conditions resulted in an effective performance by the PR.

Cost of production

Madagascar has no facilities for fertilizer production, and investment of capital and management staff in a small factory is not profitable. It is better to rent out the services of an existing dolomite factory at Antsirabe for grinding and bagging the PR. Hence, the cost of production will include the mining of PR from Andrano Island, transport by beacher (boat landing directly on the beach) to the port of Morondava, return transport of soil to Andrano, transport by truck from Morondava to Antsirabe, and grinding and bagging at the factory. The estimated ex-factory cost is US$102/tonne (US$0.51/kg P_2O_5) compared with US$418/tonne (US$0.92/kg P_2O_5) for TSP. These figures highlight the cost-effectiveness of the local Andrano PR compared with imported water-soluble P fertilizer.

India

In 1990/91, India imported fertilizer raw materials at a cost of US$338 million and manufactured fertilizers to the value of US$608 million (Srinivasan, 1994). Most of the fertilizer used in the country is in the form of N, and growth in the consumption of phosphate fertilizers has not kept pace with that of N fertilizers. Consequently, there seems to be a nutrient use imbalance. It is estimated that 46 percent of Indian soils are low in available P, 52 percent are of medium P status and 2 percent are of high P status (Tandon, 1987). Thus, there is a need to increase P fertilizer application in order to achieve higher productivity.

India imports about 70 percent of the PR necessary for phosphate fertilizer production and all of the elemental sulphur (S), mainly for use in the phosphate industry (Tandon, 1991). There are sizeable PR deposits in different parts of the country including: Mussoorie PR (Uttar Pradesh), Purulia PR (West Bengal), Jhabua PR (Madya Pradesh), Singhbhum PR (Bihar) and Kasipatnam PR (Andra Pradesh). Although the total estimated reserve is 130 million tonnes, about 60 percent of the deposits are of low grade and unsuitable for the manufacture of single or triple superphosphate (Jaggi, 1986). Of the Indian PRs, the deposit at Mussoorie (estimated reserve 45 million tonnes) and possibly that at Purulia (10 million tonnes) are considered useful for direct application. A low-reactive igneous PR from Rajasthan State (Jhamar-kotra PR) that is probably not suitable for direct application is also being marketed by Rajasthan State Mines and Minerals Ltd. under the product name Raji Phos. The estimated total PR reserve is 77 million tonnes with P contents ranging from 5 to 16 percent. In 1998, the PR applied directly amounted to about 11 000 tonnes of P, or 0.6 percent of the total P consumption of 1.8 million tonnes of P.

In India, acid soils occupy an estimated 49 million ha of agricultural land. Soil pH is less than 5.5 in 29 million ha of this land, with values of 5.6–6.5 in the rest of the area (Tandon, 1987). Almost 70 percent of India's cropland is rainfed and may not be suitable for the direct application of PR. As a first approximation, assuming a P requirement of 30 kg/ha/year, then the demand for PR-phosphorus in soils with a pH of less than 5.5 is 234 000 tonnes of P per year, or 2.6 million tonnes of PR per year if added in the form of local Mussoorie PR.

Numerous field experiments have shown that the agronomic effectiveness of Mussoorie PR could be equal or similar to that of soluble P fertilizers in soils with a pH of less than 5.5 with plantation, leguminous, rice and maize crops provided there is adequate soil water (Tandon, 1987; Poojari *et al.*, 1988). In soils with higher pH values, PRs may need to be applied as partially acidulated phosphate rock (PAPR) (Basak *et al.*, 1988; Chien and Hammond 1988) (Figure 13) or as mixtures with water-soluble fertilizers (Singaram *et al.*, 1995).

The most widely used Indian PR, Mussoorie PR, contains 8–9 percent total P, 1.14 percent of organic matter and 4 percent of sulphide sulphur. The CA-extractable P of finely ground material (particle size less than 0.15 mm) is less than 10 percent, which makes it about one-third as reactive as unground North Carolina PR. It is claimed that the oxidation of sulphide sulphur to sulphuric acid and the subsequent reaction of the acid on the francolite would increase the dissolution of Mussoorie PR and thus enhance its agronomic effectiveness. It has been demonstrated that oxidation of S in PR-sulphur granules increases the dissolution of PR (Rajan *et al.*, 1983). However, the PR contains 15 percent of free carbonates, which may counter the effect of oxidation of S. In addition, Mussoorie PR contains organic carbon (1.14 percent) in the mineralogical composition, which is likely to improve its internal porosity and, therefore, the dissolution of PR-phosphorus.

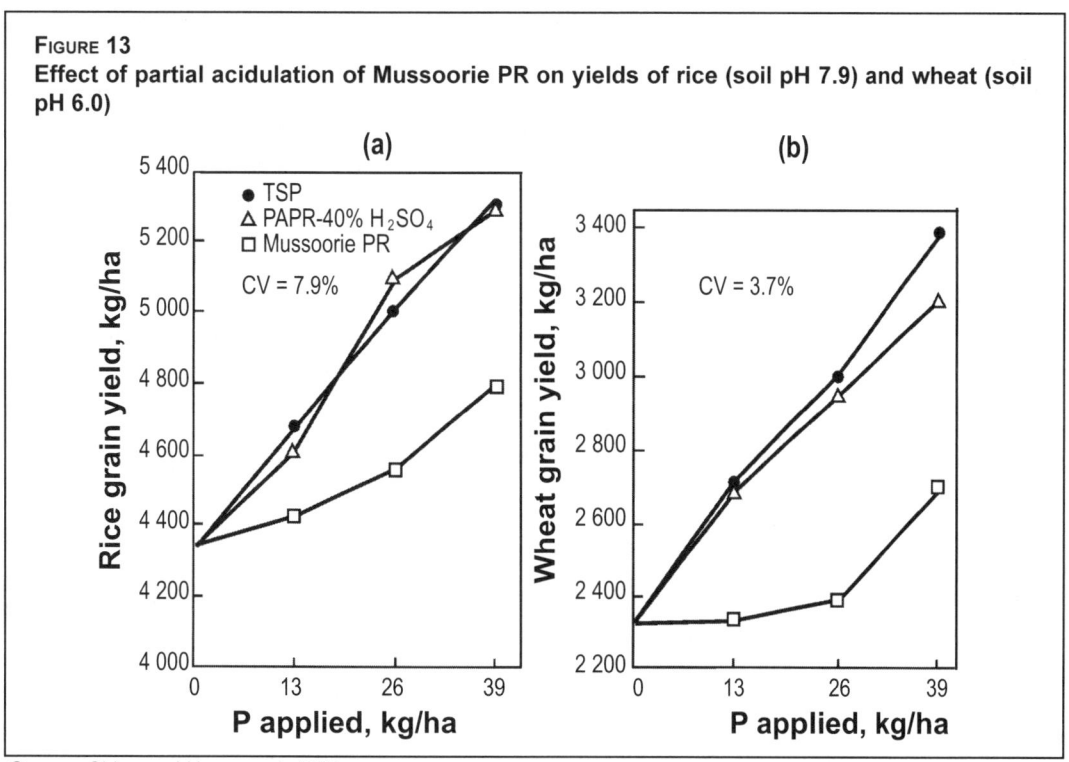

FIGURE 13
Effect of partial acidulation of Mussoorie PR on yields of rice (soil pH 7.9) and wheat (soil pH 6.0)

Source: Chien and Hammond, 1988.

Field evaluation of Mussoorie PR

Mussoorie PR was agronomically as effective as single superphosphate (SSP) in a field trial with paddy rice on an acid (pH of about 5.0) hill soil of Uttar Pradesh (Table 21) (Mishra, 1975). In another trial conducted on an acid soil of higher pH (pH 5.8), Mussoorie PR gave 95 percent of maize yield relative to SSP application whereas a 1:1 mixture was as good as SSP (PPCL, 1980). Based on the cost of Mussoorie PR being 54 percent per unit of P as that in SSP, there is a saving of 46 percent on the cost of fertilization of paddy crop.

Mussoorie PR also proved effective on soils with a high pH provided there was an adequate supply of irrigation water (Singaram *et al.*, 1995). This conclusion is based on an experiment conducted over three seasons to study the response of crops to current application as well as the residual effect of the P fertilizers. The experimental soil was a calcareous, reddish-brown, clay loam (Typic Ustropept) containing kaolinite and montmorillonite clays. The soil pH in water was 8.02, and it contained less than 1 percent of organic matter. Treatments included SSP, Mussoorie PR and a physical blend of SSP and Mussoorie PR applied at three rates of application, and a nil-P control. Three crops were grown in succession: finger millet, maize and black gram. Fertilizers were applied for finger millet and maize but not for black gram. All crops received irrigation as required.

TABLE 21
Yield of rice and maize with Mussoorie PR, SSP and a 1:1 mixture of Mussoorie PR-SSP

Treatments	Rice grain yield (tonnes/ha)				Maize yield (tonnes/ha)			
	Rates (kg P/ha)							
	0	17.5	35.0	52.0	0	17.5	35.0	52.0
SSP	2.0	2.2	2.6	2.8	5.2	6.6	9.0	10.9
Mussoorie PR		2.3	2.5	2.9		6.3	8.2	10.5
Mussoorie PR-SSP (1:1)						7.1	9.6	10.7

From the yield response curves, the substitution values (SVs) of the test fertilizers were calculated as the amount of total P applied as SSP that is required to produce 90 percent of maximum yield, divided by total P of the test fertilizer that is required to produce the same yield. Thus, a ratio of less than one indicates that the test fertilizer is less effective than SSP. The fertilizer substitution values for finger millet were 0.42 for Mussoorie PR and 0.68 for Mussoorie PR-SSP. Thus, the P supplying value of Mussoorie PR was 42 percent that of SSP. However, on application of fertilizers for maize, the respective values were 1.25 and 1.39. The residual effect of previously unreacted PR from the first application may have contributed to the substitution value exceeding one. Black-gram yields at the P application rate that gave 90 percent of maximum yield with a fresh application of SSP were 0.74 tonnes for the original SSP application, 0.74 tonnes for Mussoorie PR-SSP and 0.82 tonnes for Mussoorie PR.

TABLE 22
Net returns on P fertilizers to finger millet and maize, and black gram grown on the residual effect

Fertilizers	Finger millet	Maize	Black gram	Total
	net return (US$/ha)			
SSP	438	487	315	1 240
Mussoorie PR + SSP	438	495	316	1 249
Mussoorie PR	435	496	346	1 277

Economic analysis (Table 22) shows that the net returns for the application of Mussoorie PR and Mussoorie PR-SSP were marginally better than for SSP (Singaram *et al.*, 1995).

The above calculations do not take into account the nutritional value of the sulphate component of SSP. The value attributed to S will depend on the sulphur status of soils. Not all soils need S application. Application of PR for the first time may result in a yield penalty for the first crop. Thereafter, with applications on following crops there may be greater ongoing savings. However, the financial benefits are not particularly large for the farmers. Nevertheless, the use of local PRs can result in a substantial financial benefit for the country as a whole. This results from the saving of foreign currency as the local PRs act as import replacements. Therefore, there may need to be a subsidy to encourage farmers to use PRs.

Indonesia

Indonesia has long applied PR to plantation crops such as rubber, oil-palm, coconut, coffee, cocoa and tea. The estimated demand is 500 000 tonnes/year but the consumption is about 50 000–100 000 tonnes/year, with most of the PR coming from Jordan, Tunisia, Algeria, Morocco and Christmas Island.

Local PR resources consist of numerous cave deposits of phosphatic guano and phosphatized limestone. The Directorate of Mineral Resources estimated the total reserves at 700 000 tonnes (Harjanto, 1986). The quality of these PRs is very good with the total P_2O_5 concentration ranging from 28 to 39 percent and the FA-extractable P ranging from 54–80 percent of total P_2O_5. Therefore, these PRs are suitable for direct application.

A consortium studied the phosphate deposits in the Caimis and Tuban regions in West Java (BPPT-BRGM-CIRAD-TECHNIFERT-SPIE BATIGNOLLES, 1989). The consortium reported PR reserves of an estimated 3 million tonnes in the Ciamis region of West Java. In fact, there are many newly formed types of deposits that are solubilized and recrystalized in limestone. The phosphates also occur as impregnations of crandallite and whittokite in clay soils, which are erosion products of the same limestone. Samples from this survey have been used in pot and field experiments.

Performance of the Ciamis PRs

A pot experiment was conducted to compare three PRs (Malang, Bluri and Senori) from the Ciamis region with TSP as the reference fertilizer, all applied at 100 mg of P per kilogram of soil. The treatments included a control without P. The soil was a Podzolic soil from Singkut, Sumatra, with a pH of 4.5. There were five replications. Two test plants (*Agrostis* and soybean) were grown sequentially in pots containing 150 g of soil in order to measure the direct and residual effects respectively. The RAEs (RAE = [(yield from PR - yield from control)/(yield from TSP - yield from control)] x 100) of the three PRs were all in excess of 100 for the *Agrostis* yields while the Malang and Bluri PRs had RAEs in excess of 100 for the soybean yields. A separate incubation study was also performed with the same PRs in the same soil. The PRs increased available P to the same level as TSP. They also increased the exchangeable Ca and depressed the exchangeable Al to a greater extent than TSP. Thus, the Ciamis PRs were as effective as TSP in these studies.

A two-year field trial with five consecutive field crops was carried out on an abandoned Ultisol (pH 4.3) in South Kalimantan. The objective was to evaluate the performance of Ciamis PR against PR sources from North and West Africa, using a water-soluble local superphosphate (SP 36) as a reference fertilizer (Sri Adiningsih and Nassir, 2001). The P fertilizers were applied at a single rate of 300 kg of P_2O_5 per hectare and a control without P was included. The Ciamis PR and the highly reactive Gafsa PR from Tunisia were similar in terms of agronomic effectiveness to the water-soluble P fertilizer for the first crop and for the following four crops (Table 23). The performance of the other PRs improved with successive crops.

A simple economic analysis indicated that a large application of P fertilizer increased the net income for the farmer. The increase was US$1 050 for the water-soluble superphosphate, and US$1 264 for the ground PRs (Sri Adiningsih and Nassir, 2001). This demonstrates how the local Ciamis PRs can be used advantageously for food crop production, particularly in the reclaiming of degraded soils abandoned by the farmers.

A pilot plant was set up at Ciamis with a capacity of 10 000 tonnes/year. In 1999, the estimated production cost including mining, grinding, drying and bagging was US$50/tonne of commercial product containing 25 percent P_2O_5 (US$0.20/kg P_2O_5). The market prices of water-soluble phosphate (SP 36 with 36 percent P_2O_5) and imported PR (Gafsa with 31 percent P_2O_5) were US$171 and 114/tonne (US$0.47 and 0.38/kg P_2O_5), respectively. A joint venture has been agreed with a Japanese company to initiate exploitation of the mine. The demand is increasing but the problem is to ensure a regular supply of the commercial product.

TABLE 23
Relative agronomic effectiveness of PRs for crops on a Typic Hapludult, Pelaihari, Kalimantan

PR sources	Maize 1st crop	Upland rice 2nd crop	Cowpea 3rd crop	Maize 4th crop	Upland rice 5th crop	Average 5 crops
OCP-PR, Morocco	47	104	150	121	128	110
Gafsa-PR, Tunisia	114	105	162	113	108	119
Djebel-Onk, Algeria	25	98	162	130	133	109
ICS-PR, Senegal	69	99	112	118	95	98
OTP-PR, Togo	41	89	50	130	120	86
Ciamis-PR, Indonesia	106	114	212	90	122	128

New Zealand

Pastoral farming based on ryegrass/white-clover permanent pastures constitutes about 90 percent of agriculture in New Zealand for the production of dairy, sheep, beef and deer products. Whereas the pasture plants rely mainly on the atmospheric N fixed by clover plants, P is applied as fertilizers. Permanent pastures are particularly suited for the use of PRs because: (i) they need a steady supply of P over a long period; (ii) they possess a high root density; and (iii) legumes (clover in this case) are efficient users of PR because of their affinity for Ca and the acidifying effect of N fixation on rhizosphere. In addition, most New Zealand soils are slightly acidic (pH 5–6) with adequate moisture regimes (more than 850 mm rainfall per year), both of which favour PR dissolution (Hedley and Bolan, 2003). About 10 percent of P fertilizers are applied as reactive PRs (about 142 000 tonnes of RPR), and PRs are reported to be gaining about 1 percent of the phosphate market each year (Quin and Scott, 2003). As PR application is allowed under 'organic farming' practices, there is an incentive to use PR in some situations.

The first series of trials to evaluate PRs as P fertilizers began in 1932 (Hedley and Bolan, 1997). Since the mid-1970s, systematic studies have included numerous field trials, greenhouse experiments and laboratory tests (Hedley and Bolan, 2003).

The results from the field trials show that there might be a delay before PRs begin to be effective, and it may take 4–6 annual applications before reactive PRs become as effective as TSP. In these trials, the PRs were surface applied and not incorporated into the soil. The time delay can be attributed to the time required for the PR to be incorporated into the soil through the effect of worm activity, rainfall, etc. and for PR-phosphorus to dissolve. Moreover, as the concentration of P around PR particles is about 4 mg/litre (Watkinson, 1994b), it will take longer than for SSP for the flux of P into the bulk soil for absorption of P by plants. Studies have shown that even after four annual applications of North Carolina PR, P concentration in solution does not increase significantly below 20 mm soil depth in a highly P retentive soil (Rajan, 2002).

The decline in absolute dry-matter production in the first few years could be substantial in soils of low P fertility. However, in soils of above-medium fertility, the decline may be less than 10 percent (Quin and Scott, 2003). This is because the increase in pasture production resulting from P fertilizer application can be expected to be small in these sites. In New Zealand, RPR is applied increasingly with water-soluble P in order to enhance its agronomic effectiveness (Chapter 9).

Currently, one kilogram of P in RPR costs US$0.87, whereas it costs US$1.07 as SSP if the soil does not require S. Thus, RPR phosphate is 20 percent cheaper. A promising strategy is to encourage application of RPR with soluble P, which is likely to result in the same level of agronomic effectiveness as the application of soluble P (Chapter 9). The savings on using RPR will be less if the value of S in the fertilizers is accounted for. Thus, the cost-effectiveness of RPR relative to soluble fertilizers will depend on the S requirement of soils in addition to environmental and management factors (Sinclair *et al.*, 1993a; Rajan *et al.*, 1996). A specific incentive for RPR use in New Zealand is its permissibility for use in organic farming.

Latin America

Direct application of PR has received much attention in Latin America in the past 20 years. Experiments have evaluated the agronomic and economic potential of indigenous PRs found in each country. The main objective has been to determine whether the local PRs could be used after grinding, or modified to produce PAPR, or subjected to high temperatures to produce thermophosphate, in order to reduce countries' dependency on imported water-soluble P fertilizers (Casanova, 1995). The results have shown that PR use is advisable from some sources and under certain conditions (Casanova, 1998; Lopes, 1998; Besoain et al., 1999).

The savannahs located in the tropics and subtropics are the main agricultural frontier of Latin America. The soils, mainly Ultisols and Oxisols, are highly weathered, of low fertility, very acidic and with a high P-fixing capacity (Van Uexkull and Mutert, 1995). Therefore, the use of P fertilizer is important to maintaining and increasing agricultural productivity in these soils (Casanova, 1992). This situation, together with the large PR reserves (especially in Venezuela and Brazil) and the high costs of imported P fertilizers, has promoted a diversification in the production of P fertilizers using PR sources in these countries (Casanova, 1995). Several strategies for a more rational use of P fertilizers have been proposed. The following case studies on Venezuela and Brazil illustrate how local PRs can be modified and used effectively by farmers.

Venezuela

With reserves of 2 650 million tonnes, Venezuela has the third largest PR reserves in Latin America (after Mexico and Peru). The total P_2O_5 content ranges from 20 to 27 percent and the P_2O_5 solubility in 2-percent CA ranges from low to medium, using the solubility criteria reported by Hammond and Leon (1983).

Management guidelines for the direct application of ground or modified PRs have been developed over the last 20 years for annual and permanent crops for different soil and climate conditions (Casanova, 1992; Casanova et al., 1993). The P requirements for these crops, together with the methods and rates of application and their residual effects, have been determined using laboratory, greenhouse, field experiments and commercial evaluations of these P sources (Casanova et al., 2002b). This work has provided the basis for the opening of a fertilizer plant for producing granular PAPR with a capacity of 150 000 tonnes/year. The PR is partially acidulated with locally produced 60–75-percent sulphuric acid (Chapter 10).

A number of factors support the use of PAPR from such a plant in Venezuela. About 75 percent of the soils in the country are acid and the climate conditions are mainly tropical with an average rainfall of about 1 000 mm/year. There are 1 million ha of annual crops and 6.5 million ha of permanent crops (Comerma and Paredes, 1978).

A recent commercial evaluation of the agronomic and economic results from the use of PAPR in the rehabilitation of degraded pastures in Monagas State has found this approach to be very successful. The evaluation involved 30 farms using a dual-purpose milk and meat production system based on highly degraded *Brachiaria brizantha* pastures. The pasture degradation resulted from overgrazing and low fertilizer input on soils that were very deficient in N and P. The degraded pasture grew to heights of 20–25 cm and covered 30–40 percent of the soil surface. The dry-matter yields were 300 kg/ha/harvest and this forage contained 2–3 percent of crude protein and was of low digestibility. The average milk production per animal was 3–5 litres/d.

These degraded pastures were rehabilitated with a per-hectare application of 200 kg of PAPR and 100 kg of urea mixed with 3 kg/ha of *Stylosanthes capitata* seed at the beginning of the rainy season. Table 24 shows the results obtained with this approach. The crude-protein concentration of the pasture increased to 6 percent, the P, Ca, Mg, iron (Fe), copper (Cu), zinc (Zn) and manganese (Mn) concentrations increased to sufficiency levels, while per-animal milk production increased to 8 litres/day in the 100-animal herd. After one year, the animals had increased in weight from 350 kg on the traditionally managed degraded pastures to 470 kg for the PAPR-urea-*Stylosanthes* management system. The increased profits were an additional US$90/day from the milk production and US$15 600 in meat production for the 100-animal herd.

TABLE 24
Comparison between a rehabilitated pasture using the PAPR-urea-*Stylosanthes capitata* treatment and a traditional degraded pasture, Monagas State, Venezuela

Parameter measured	PAPR-treated, rehabilitated pasture	Degraded pasture
Pasture yield (tonnes DM/harvest)	1.8	0.3
Pasture composition (DM basis)		
Crude protein (%)	6.4	3.0
P (%)	0.15	0.08
K (%)	1.02	0.50
Ca (%)	0.26	0.20
Mg (%)	0.46	0.25
Fe (mg/kg)	332	214
Cu (mg/kg)	5	1
Zn (mg/kg)	45	37
Mn (mg/kg)	154	83
Milk production (litres/animal/day)	8	5
Rehabilitation cost (US$/ha)	77	462

Rehabilitating the degraded pasture by the traditional method using soil tillage, fertilization with soluble fertilizer, herbicides, planting new seed, and a delay until the pasture covered all the soil, cost US$462/ha compared with US$77/ha for the PAPR treatment (Table 24). Thus, all 33 farmers increased their pasture, milk, meat and profitability in a sustainable production system with the PAPR technology.

The social impact is that farmers now have phosphate fertilizer available in more than 175 fertilizer dealers across the country, at a competitive price of US$49/tonne compared with the highly water-soluble, imported TSP (US$148/tonne). Both fertilizers are similar in terms of agronomic and economic efficiency.

Brazil

Brazil has PR reserves of 376 million tonnes, which have a total P_2O_5 content range of 24–38 percent. However, the P_2O_5 solubility in 2-percent CA for the mainly igneous PRs is low (2.6–4.8 percent). Brazil uses about 4 600 tonnes of P per year (almost 3 percent of world consumption). The approaches taken in Brazil have been to either import water-soluble P fertilizers or reactive PRs, or treat indigenous PRs at high temperature in a mixture with basic slags in order to produce thermophosphate (Table 25).

TABLE 25
AEI of phosphate fertilizers in a clayey Oxisol of central Brazil, based on P uptake data over five years with annual crops, followed by three years with *Andropogon* pasture

Phosphate fertilizer	200 kg P_2O_5/ha			800 kg P_2O_5/ha		
	Annual crop	Andropogon	Total	Annual crop	Andropogon	Total
TSP*	100	100	100	100	100	100
Gafsa	93	110	104	106	106	106
Thermophosphate Mg	92	142	113	110	119	114
Thermophosphate IPT	45	84	60	88	98	92
Araxa**	27	69	41	47	74	58
Patos**	45	81	59	56	91	70
Catalao**	8	36	17	26	43	33

*AEI% = [(crop yield using fertilizer (PR) - crop yield of control plot)/(crop yield of reference fertilizer (TSP) - crop yield of control plot)] x 100.
** Brazilian PRs.

A study by Lopes (1998) reports an experiment conducted on a clayey Oxisol with a pH of 4.2 that had received 2.2 tonnes of lime per hectare. The different P fertilizers, including local PRs (Araxa, Patos and Catalao), were evaluated at 200 and 800 kg of P_2O_5 per hectare and were broadcast in the first year. The results (Table 25) show that the Brazilian PRs have a low agronomic efficiency index (AEI). In five years of annual crops, the AEI was less than 50 percent for all PRs at both rates of P_2O_5 except for Patos PR at 800 kg/ha. With *Andropogon* grass, grown for three years after the annual crops, the AEI was considerable higher, except for Catalao PR. The high AEI for the Gafsa PR is a good indication that this product is an effective source of P in these acid soils.

Lopes (1998) also describes technologies, developed from a series of laboratory studies, glasshouse work and field experiments, that have brought millions of hectares of unproductive land into highly productive farmland. The best strategies developed to build up capital P in these soils are:

- In order to achieve adequate yields in four years, a P application of 1.5–2 kg/ha for each 1 percent of clay is broadcast as water-soluble P fertilizers, thermophosphate or highly reactive PR according to crop needs. This strategy was widely adopted and resulted in large areas of land being brought into crop production in the 1970s and 1980s. The strategy was combined with subsidies and adequate credit funding at low interest rates. The approach is in use today, without subsidies.

- In order to achieve adequate yields in six years, annual applications of mainly soluble P fertilizers, but also thermophosphate and highly reactive PR, are banded according to crops needs. In addition, a small excess (8–10 kg of P per hectare) is applied every year. The subsequent tillage operations and the residual effect of fertilizer P lead to a gradual build-up of P. Resource-poor farmers with insufficient funding for large amounts of P as a capital investment have tended to use this option, which is more consistent with the present agricultural policy in Brazil.

The study by Lopes (1998) points out that these management strategies, developed in the 'Cerrado' region of Brazil, have resulted in millions of hectares being developed for crop-livestock production systems. It further remarks that these results demonstrate that the highly weathered, acid, nutrient-depleted soils of the tropics can be as productive as the better soils of the world with the use of this appropriate technology.

Chapter 6
Soil testing for phosphate rock application

Soil testing for phosphorus (P) has been a subject for extensive research. Numerous extractants ranging from strong acids to alkalis and various organic and inorganic complexing agents have been developed to evaluate P bioavailability with certain crops and soils. The most widely used soil P tests are Bray I (Bray and Kurtz, 1945), Mehlich I (Nelson *et al.*, 1953), and Olsen (Olsen *et al.*, 1954). Other common tests include Bray II, Mehlich II and III, and resin (Fixen and Grove, 1990). However, all these soil tests are mainly for recommendations with water-soluble P fertilizers such as di-ammonium phosphate (DAP), single superphosphate (SSP) and triple superphosphate (TSP). Reports have shown that these conventional acid or alkaline soil tests do not work well in soils fertilized with phosphate rock (PR) (Perrott *et al.*, 1993; Menon and Chien, 1995; Rajan *et al.*, 1996). Thus, there is a need to develop appropriate soil tests that reflect closely P uptake from PR over a wide range of PR sources, soil properties, and crop varieties. Furthermore, the soil tests should be suitable for PR and for water-soluble P fertilizers. This issue has become more important because of the increasing interest in the use of PR for direct application in developed and developing countries, e.g. Australia, New Zealand, Brazil, Indonesia, Malaysia and countries in sub-Saharan Africa.

This chapter provides an overview of the soil tests that have been used to measure available P from soils treated with PR and water-soluble P fertilizer. It provides an introduction to the nature of the reactions of PR and water-soluble P fertilizer in soils. It then discusses the use of the ^{32}P isotopic exchange kinetic method in relation to soil-available P measurement.

SOIL PHOSPHORUS REACTION AND SOIL TESTING

When a water-soluble P fertilizer, e.g. TSP, is applied to acid soil containing oxides of iron (Fe) and aluminium (Al), reaction products in the form of Fe-Al-P are the sources of available P through desorption/dissolution processes for uptake by the plant (Figure 14). If the amount of available P extracted by a soil test (k_1') is proportional (i.e. good correlation) to the amount of P absorbed by the plant (k_1) or the crop yield, then this soil test is suitable for calibration for fertilizer-rate recommendations of water-soluble P.

When a PR is applied to an acid soil (Figure 14), dissolution of the PR releases P to the soil solution. The plant roots can absorb directly part of the released P, and part of the released P forms reaction products (above). These can also provide available P later through the desorption process (unlike water-soluble P, soil solution P concentration from PR dissolution is too low to precipitate Fe-Al-P compounds). Thus, both the reaction products and undissolved PR can provide available P to the plant (Chien, 1978). The degree of relative P availability from undissolved PR versus Fe-Al-P depends on: PR reactivity, soil P-fixing capacity, and the duration of the PR reaction with soil. Although the reaction products of PR dissolution may be assumed to be the same as that of TSP dissolution, i.e. $k_3/k_3' = k_1/k_1'$, the P mineral in PR is calcium (Ca) P.

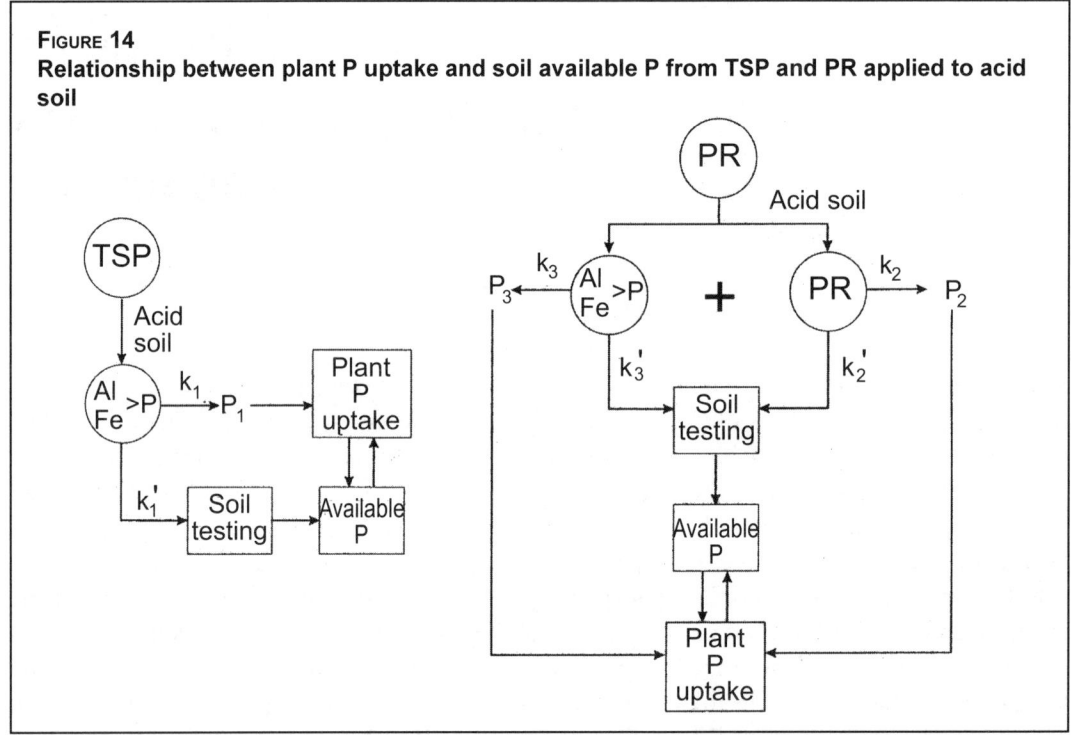

FIGURE 14
Relationship between plant P uptake and soil available P from TSP and PR applied to acid soil

This differs from the reaction products of TSP in the form of Fe-Al-P. Thus, the proportionality of k_1/k_1' for TSP may or may not be the same as that of k_2/k_2' for PR with a given soil test. If $k_2/k_2' > k_1/k_1'$, the soil test will underestimate the available P from PR with respect to TSP. If $k_2/k_2' < k_1/k_1'$, the soil test will overestimate the available P from PR with respect to TSP. If $k_2/k_2' = k_1/k_1'$, then the soil test will be applicable to both PR and TSP.

As Figure 14 shows, as long as the undissolved PR is the primary source for providing available P to the plant, there will be two separate calibration curves with a given soil test (one for PR and one for TSP) if $k_2/k_2' \neq k_1/k_1'$. When most of the PR is dissolved and the reaction products of PR are the main sources of available P to the plant, both TSP and PR will follow the same calibration curve if $k_3/k_3' = k_1/k_1'$ within a given soil test. However, there is some evidence to suggest that the reaction products of PR with soil and their availability to plants may differ from those obtained with soluble P fertilizers, i.e. $k_3/k_3' \neq k_1/k_1'$, owing to the slow-release nature of PR dissolution (Chien *et al.*, 1987a). Therefore, whether a soil test is applicable to both TSP and PR depends on: the chemical nature of the test (acidic, alkaline, etc.); the amount of undissolved PR; and the amount and nature of the reaction products formed from PR dissolution.

CONVENTIONAL SOIL TESTS

Bray I test

The Bray I solution (0.03 M NH_4F + 0.025 M HCl) developed by Bray and Kurtz (1945) has been employed widely to determine the available P in soils. It has given results that are highly correlated with crop response to P fertilization. The combination of weak HCl and NH_4F is designed to remove soluble forms of P by the acid and precipitation of CaF_2, largely Ca-P other than apatite, and a portion of Fe-Al-P by complexing cations with fluorine (F) ions. Therefore, many researchers have assumed that the P extracted by Bray I from soils treated with PRs is the available P derived from the reaction products of PRs rather than from the undissolved PRs in the soils. However, Chien (1978) found that Bray I was able to dissolve PRs in an acid soil at time zero, i.e. before the soil reacted with PRs. The amounts of Bray I-P extracted from the soil treated with PRs before and after incubation (three weeks) correlated well with PR solubility and P rate. Thus, Bray I can help predict crop response to various sources of PR applied at different P rates without comparing with water-soluble P fertilizers. Numerous reports support this statement (Barnes and Kamprath, 1975; Smith *et al.*, 1957; Smith and Grava, 1958; Howeler and Woodruff, 1968; Hammond, 1977; Mostara and Datta, 1971). However, the question arises as to whether Bray I can be used to compare available P in soil treated with PR to that with water-soluble P fertilizers in terms of crop response to P.

Barnes and Kamprath (1975) reported that two distinct curves were obtained with a Bray I-P level from two PRs and TSP plotted against the dry-matter yield of maize. These curves indicated that the dry-matter yield of maize with a given Bray I-P level in a PR-treated soil was higher than the yield from the same Bray I-P level in the TSP-treated soil. Furthermore, Bray I-P extracted from the soils treated with PRs (Florida and North Carolina, the United States of America) correlated closely with plant yield and P uptake regardless of the changes in soil types, soil pH and source of PR. However, Barnes and Kamprath attempted to explain why two distinct curves were obtained when they plotted dry-matter yield of maize versus Bray I-P obtained with PRs and TSP. One explanation was that there are some 'acidulation products' from PR that the plant can utilize but the extractant cannot dissolve.

Figure 15 shows two more examples of the use of Bray I to evaluate available P of PRs as compared with TSP. These results show that the relationship between dry-matter yield and Bray I-P for various PR sources was the same, i.e. a single curve fitted all PR sources. Furthermore, the results show that the Bray I-PR curve was above the TSP curve. This suggests that Bray I underestimated available P from PRs with respect to TSP because a higher dry-matter yield was obtained with PRs than TSP at the same level of extractable Bray I. Apparently, the weak acidic Bray-I solutions extracted less P from apatite in the undissolved PR than the reactive

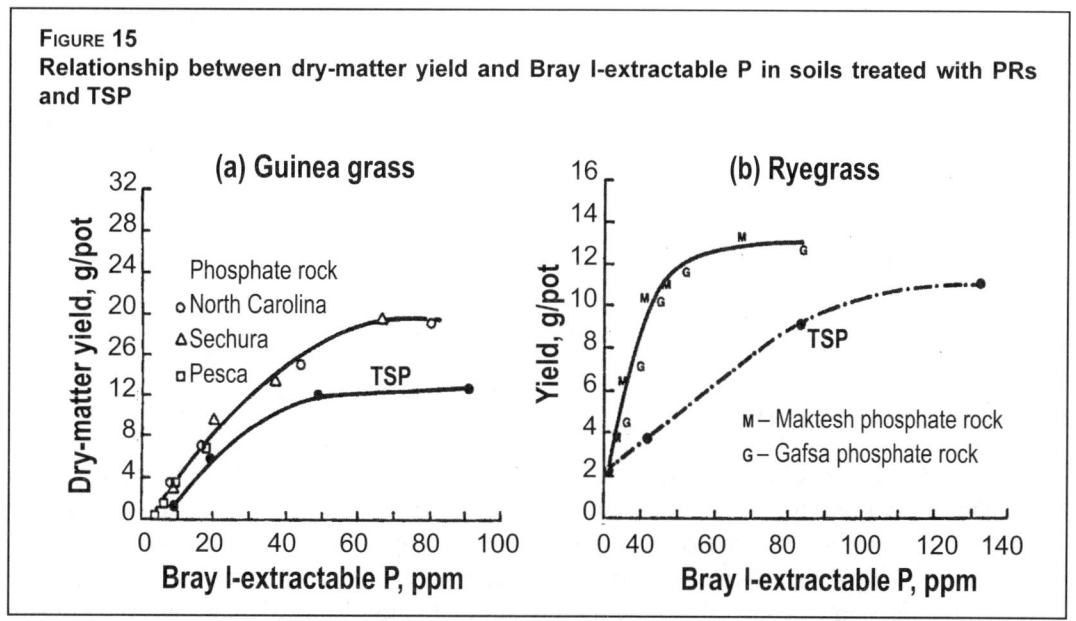

FIGURE 15
Relationship between dry-matter yield and Bray I-extractable P in soils treated with PRs and TSP

Sources: Data for (a) from Hammond (1977) and (b) from Reinhorn and Hagin (1978, unpublished data).

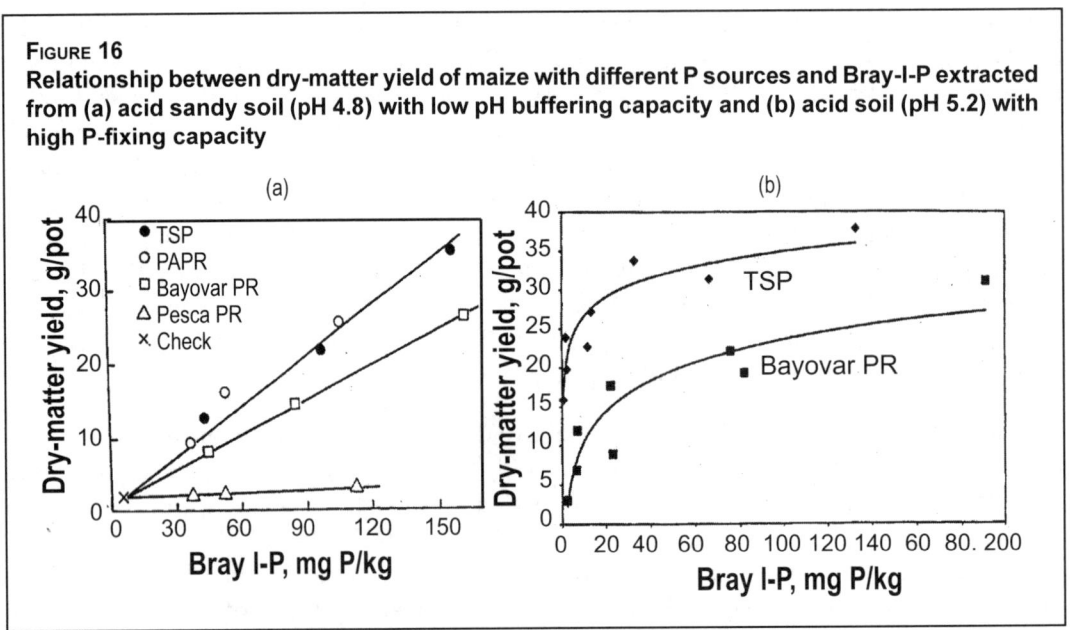

FIGURE 16
Relationship between dry-matter yield of maize with different P sources and Bray-I-P extracted from (a) acid sandy soil (pH 4.8) with low pH buffering capacity and (b) acid soil (pH 5.2) with high P-fixing capacity

Sources: Data for (a) from Chien *et al.* (1987b) and (b) from Hammond *et al.* (1986a).

products of TSP with respect to the relative P uptake by plants from PR versus TSP, i.e. k_2/k_2' > k_1/k_1' (Figure 14).

As Figure 16 shows, Bray I can also overestimate available P from PR with respect to TSP in acid sandy soil or high P-fixing soil. In sandy soils with relatively low pH buffering capacity, the pH of soil suspension can still be acidic during soil extraction. For example, the pH of Bray-I extract from the acid sandy soil (pH 4.8) shown in Figure 16(a) was 3.1, little changed from the pH of the Bray I solution (pH 3.0). Thus, the Bray-I solution can dissolve proportionally more P from PR than the reactive products of TSP with respect to P uptake from the two sources, i.e. $k_2/k_2' < k_1/k_1'$, as shown in Figure 14. In soils with relatively high pH buffering capacity, the acidity of the Bray-I solution can be neutralized partially by soil so that it is not strong enough to extract P from PR excessively. For example, the pH of Bray-I extract from an acid soil (pH 5.2) containing 34 percent clay was 4.6, higher than the pH of the Bray-I solution (pH 3.0). For these types of soil, Bray I frequently underestimates available P from PR with respect to TSP as shown in Figure 15.

In soils with relatively high P-fixing capacity, the plant-available P from PR dissolution in soil solution (i.e. k_2 in Figure 14) can be relatively low because P released from PR is fixed by the soil. However, the Bray-I solution can still extract P from PR during soil extraction (i.e. k_2' in Figure 14). Therefore, Bray I can overestimate available P from PR with respect to TSP in the soils with a high P-fixing capacity shown in Figure 16(b), i.e. $k_2/k_2' < k_1/k_1'$ in Figure 14.

Bray II and Mehlich I tests

Bray II (0.03 M NH_4F + 0.1 M HCl) and Mehlich I (0.05 M HCl + 0.0125 M H_2SO_4) are more acidic than Bray I because of relatively higher concentrations of HCl. These two soil tests are commonly used in tropical acid soils, e.g. Bray II in Malaysia and Colombia, and Mehlich I in Brazil. Because of the relatively strong acidic reagents of these two tests, they can dissolve a substantial amount of undissolved PR during soil extraction. Hence, they can overestimate

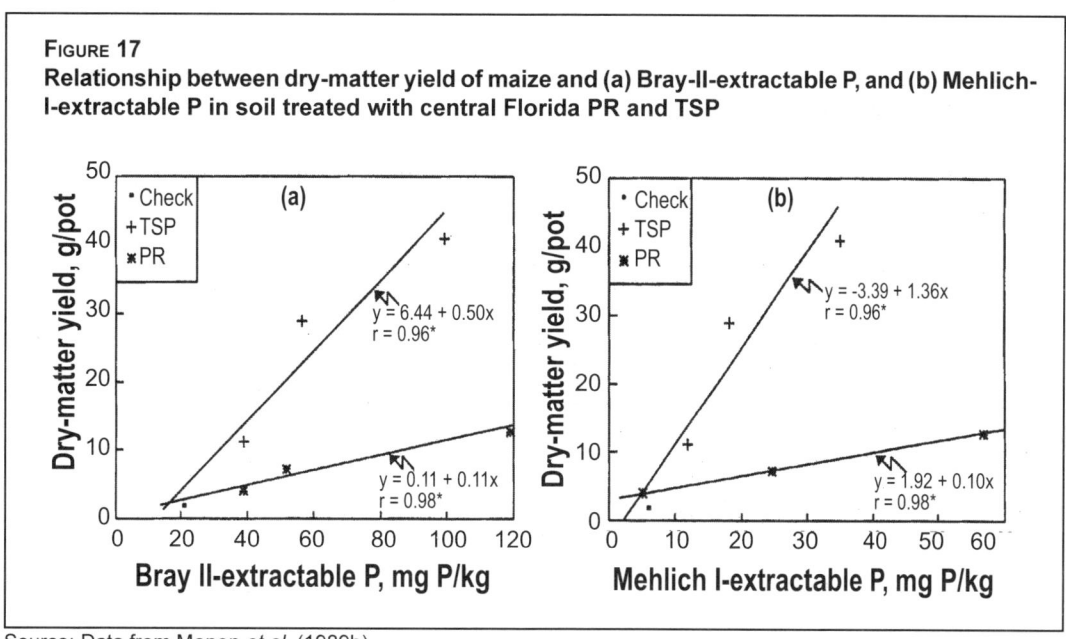

Source: Data from Menon *et al.* (1989b).

PR availability with respect to water-soluble P (Barnes and Kamprath, 1975; Yost *et al.*, 1982; Bationo *et al.*, 1991). Figure 17 shows that both Bray II and Mehlich I extracted proportionately more P from PR than the reaction products of TSP with respect to P availability, i.e. $k_2/k_2' < k_1/k_1'$, as shown in Figure 14. Therefore, the Bray II and Mehlich I methods are not recommended for PR and water-soluble P as they can overestimate available P from PR as compared with water-soluble P.

Olsen test

Olsen (0.5 M $NaHCO_3$, pH 8.5) is an alkaline reagent that extracts P from the reaction products of PR (i.e. k_3') in soil treated with PR similar to the reaction products of TSP (i.e. k_1') (Figure 14).

Although k_3/k_3' of PR may be similar to k_1/k_1' of TSP, the Olsen reagent does not extract P from undissolved PR (i.e. $k_2' = 0$). However, undissolved PR can be a more important P source than its reaction products for providing available P to the plant, especially the initial P effect. In this case, the Olsen test usually underestimates available P from PR as compared with that from TSP with respect to P uptake by the plant, i.e. $k_2/k_2' > k_1/k_1'$ (Figure 14). Therefore, the Olsen test frequently results in two curves with the PR curve above the curve of water-soluble P when crop yield or P uptake is plotted against Olsen-P (Figure 18).

In the study by Rajan *et al.* (1991b), the slope of the calibration curve for PR was very steep due to a narrow range of Olsen P (Figure 18(a)) that represented only the available P from the reaction products of PR but not from the undissolved PR. To have the same calibration curve for both PR and TSP, Perrott *et al.* (1993) used a correction factor (1.69) to multiply the Olsen P values of PR and moved the PR curve to the TSP curve (Figure 18(b)). However, this modification is an empirical approach and depends on several factors, e.g. time of PR dissolution, reactivity of PR source, soil properties, and crop species.

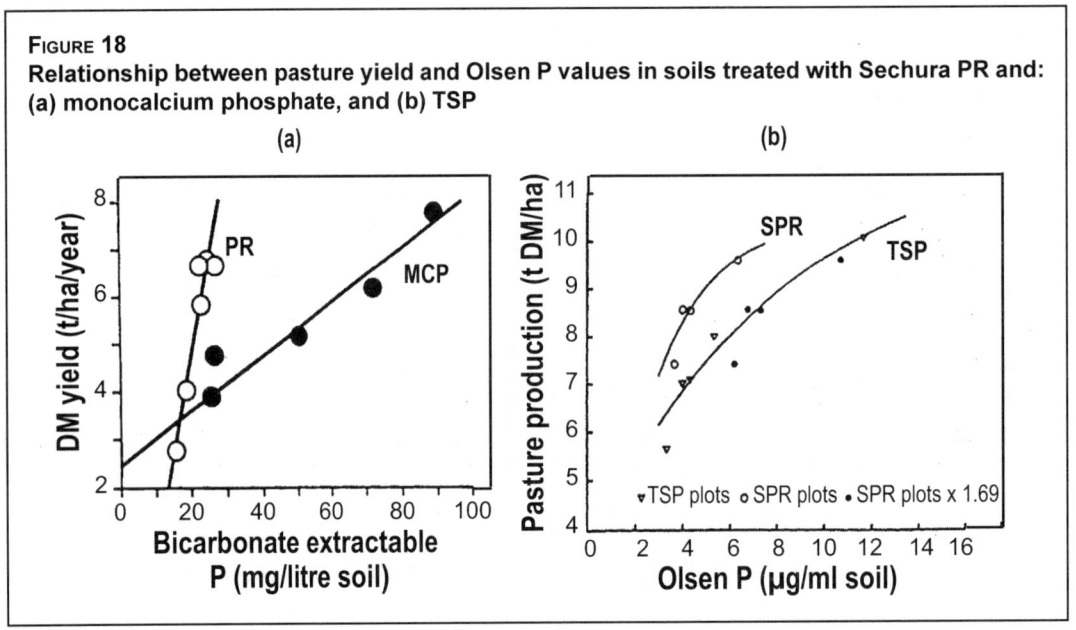

FIGURE 18
Relationship between pasture yield and Olsen P values in soils treated with Sechura PR and: (a) monocalcium phosphate, and (b) TSP

Sources: Rajan et al., 1991b; Perrott et al., 1993.

RECENTLY DEVELOPED SOIL TESTS

Iron-oxide-impregnated paper test

The iron-oxide-impregnated paper (P_i) strips test is a new approach for evaluating available P in soils (Menon et al., 1989b; Menon et al., 1990; Chardon et al., 1996; Menon et al., 1997). Instead of extracting solutions, the method uses P_i strips as a sink to sorb and retain P mobilized in a soil suspension, similar to anion exchange resin. However, unlike the anion exchange resin, the P_i strips show a preference for phosphate over other anions that are common in the soil (Van der Zee et al., 1987). Moreover, the phosphate ions can be retrieved easily from the soil suspension. Furthermore, preparation of P_i strips is sufficiently simple and inexpensive as to enable almost all laboratories to produce them as opposed to purchasing expensive anion exchange resin from chemical companies (Menon et al., 1989a). Unlike the reagents used in conventional soil tests, the P_i strips do not react with the soil but only sorb the P entering the soil solution. Thus, the method has potential for use in soils fertilized with water-soluble P and PR fertilizers.

Menon et al. (1989b) reported on the use of P_i strips for available P in soils treated with a medium-reactive Florida PR or TSP before cropping with maize. Compared with other conventional soil tests, the P_i test had the highest correlation between dry-matter yield or P uptake and soil-available P when all the soils, P sources, and P rates were pooled. However, in the study, the ranges in plant yield or P uptake and P_i-P obtained with TSP were much wider than those obtained with Florida PR. Thus, statistically, it was not appropriate to run the same regression line between crop response (plant yield or P uptake) and P_i-P for both PR and TSP as the regression was dominated by TSP (although the correlation coefficient was highly significant). In order to evaluate critically whether PR and TSP follow the same relationship when crop response is plotted against P_i-P in soil, the ranges in crop response and P_i-P with PR and TSP should be approximately the same. Furthermore, the initially developed P_i test with 0.01 M $CaCl_2$ in soil extraction could have underestimated available P from PR with respect to TSP in the study by Menon et al. (1989b) because of Ca common-ion effect on PR dissolution, i.e. $k_2/k_2' > k_1/k_1'$ (Figure 14).

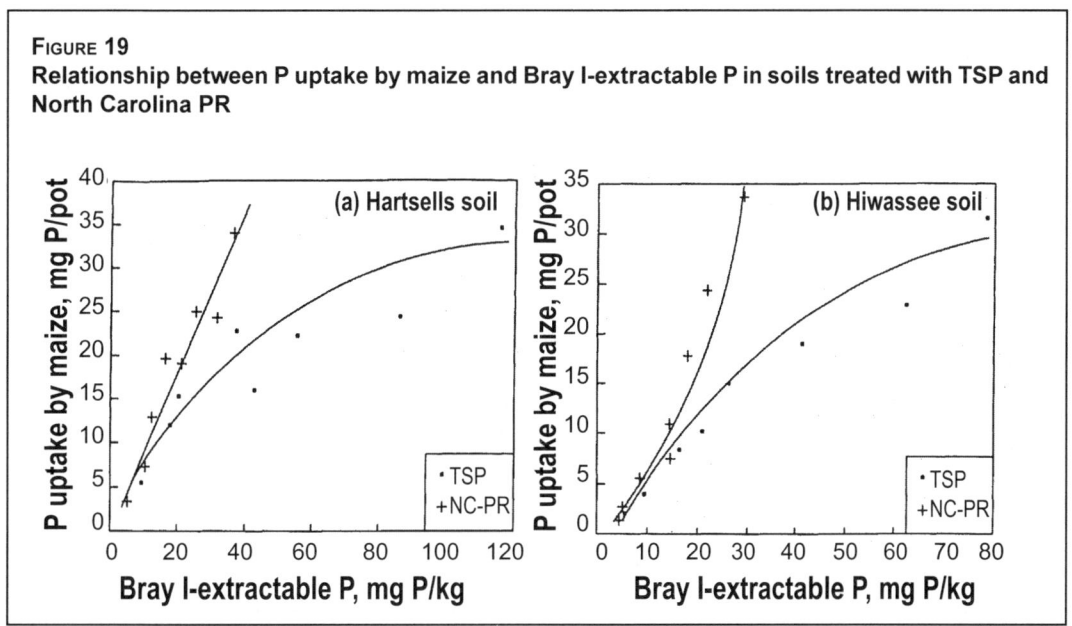

FIGURE 19
Relationship between P uptake by maize and Bray I-extractable P in soils treated with TSP and North Carolina PR

Source: Habib *et al.*, 1998.

Habib *et al.* (1998) pre-incubated two acid soils with a highly reactive North Carolina PR and TSP followed by soil extractions for available P before cropping with maize. The ranges of crop response (dry-matter yield or P uptake) obtained with North Carolina PR were within 83–91 percent of TSP. The soil tests used were P_i strips with 0.01 M $CaCl_2$, P_i strips with 0.02 M KCl, and Bray I. The results showed that both Bray I and P_i strips with 0.01 M $CaCl_2$ underestimated available P from North Carolina PR (Figures 19 and 20), confirming $k_2/k_2' > k_1/k_1'$ (Figure 14). Available P estimated by P_i strips with 0.02 M KCl instead of 0.01 M $CaCl_2$ was related more closely to crop response for soils treated with North Carolina PR and TSP (Figure 21), i.e. $k_2/k_2' = k_1/k_1'$ (Figure 14). Less P was extracted from North Carolina PR by the P_i strips with 0.01 M $CaCl_2$ than with 0.02 M KCl (Figure 22) because $CaCl_2$ depressed

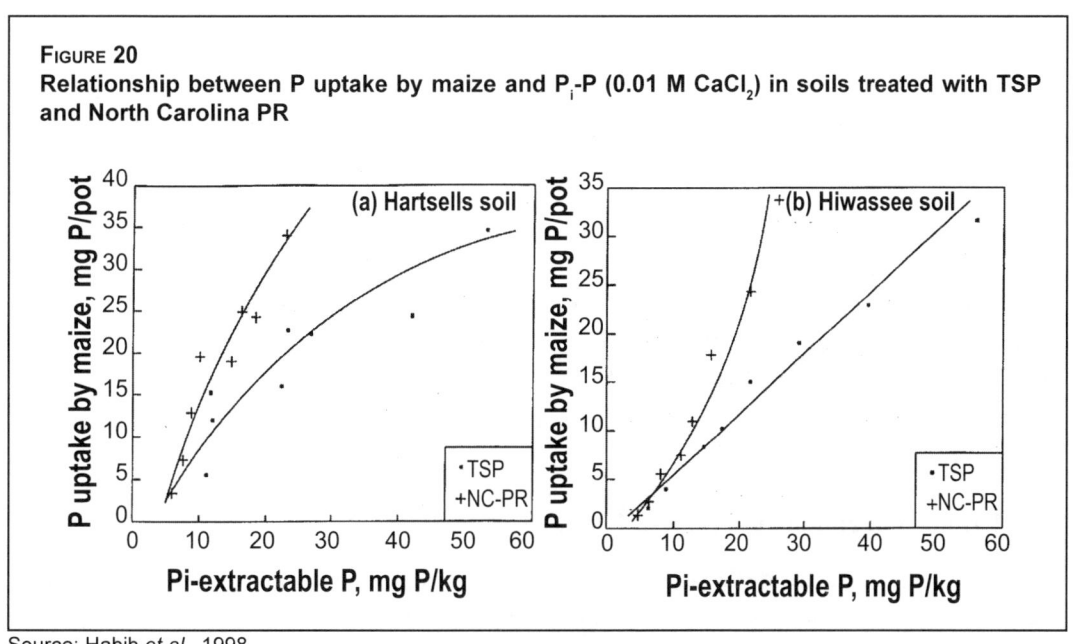

FIGURE 20
Relationship between P uptake by maize and P_i-P (0.01 M $CaCl_2$) in soils treated with TSP and North Carolina PR

Source: Habib *et al.*, 1998.

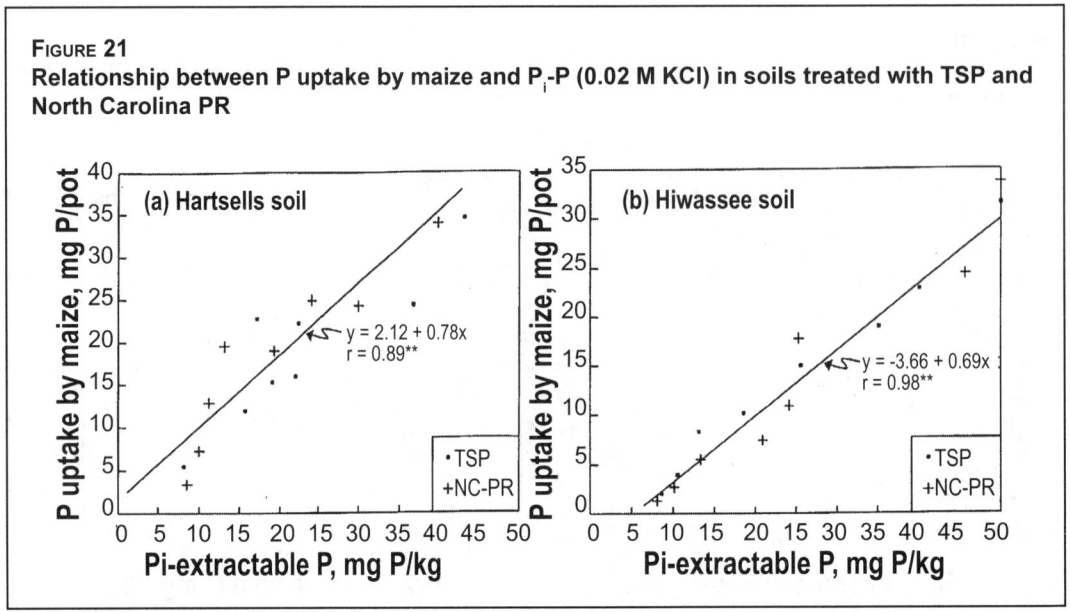

FIGURE 21
Relationship between P uptake by maize and P_i-P (0.02 M KCl) in soils treated with TSP and North Carolina PR

Source: Habib et al., 1998.

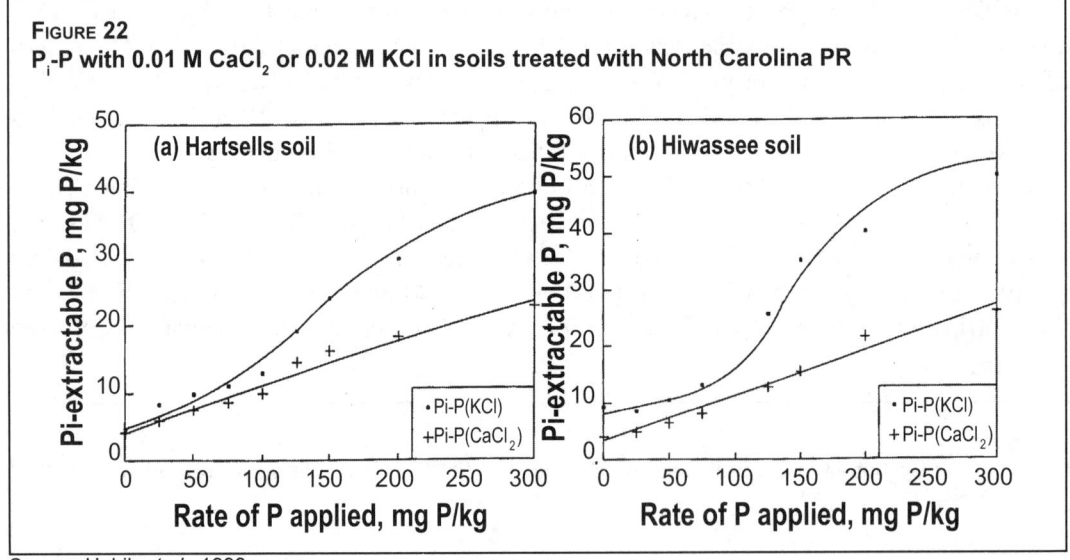

FIGURE 22
P_i-P with 0.01 M $CaCl_2$ or 0.02 M KCl in soils treated with North Carolina PR

Source: Habib et al., 1998.

PR dissolution, whereas no effect was observed with TSP (Figure 23) owing to the lack of a Ca common-ion effect on P extraction from its reaction products (Fe-Al-P). Therefore, it is recommended that the P_i test with 0.02 M KCl be used to estimate available P in soils treated with both PR and water-soluble P fertilizers. However, more research is needed, especially agronomic trials under field conditions, to test this modified P_i test.

Mixed anion and cation exchange resins

Since its introduction by Amer *et al.* (1955), anion exchange resin has been used widely in research work to evaluate soil-available P. Like the P_i test, anion resin is a non-destructive method that simulates the activity of plant roots. However, it is a time-consuming procedure involving

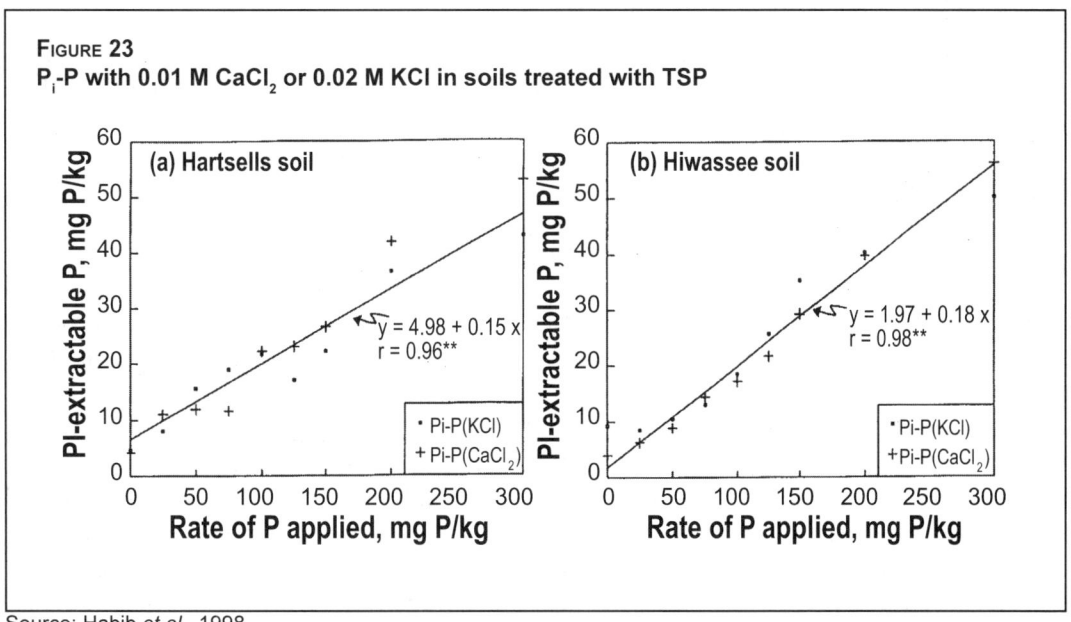

FIGURE 23
P_i-P with 0.01 M $CaCl_2$ or 0.02 M KCl in soils treated with TSP

Source: Habib et al., 1998.

a two-step anion-exchange process (resin anion exchange with soil P, followed by exchange of resin P with anion). Moreover, the difficulty of separating the small resin beads from the soil and its cost have limited the use of this method to some countries, e.g. Brazil and Denmark.

Saggar et al. (1990) introduced a simplified resin membrane strip containing a mixture of cation (Na^+) and anion (HCO_3^-) to evaluate available P in the soils treated with PR and water-soluble P. The inclusion of Na^+-saturated resin will enhance P release from PR by removing Ca^{+2} associated with PR so that it does not underestimate PR availability but will not excessively dissolve PR to overestimate PR availability, i.e. $k_2/k_2' = k_1/k_1'$ (Figure 14). Saggar et al. (1990) simplified the procedure by shaking the P-retained strip directly with the colour-developed reagent for P and, thus, they omitted an elution step (traditional exchange of P-retained by resin with NaCl solution). The amounts of P extracted by the mixed resin procedure were in proportion to the amount and solubility of the P source applied to the soils. Subsequently, Saggar et al. (1992a, 1992b) compared the mixed cation-anion resin method with seven other soil tests and found the mixed method was best for assessing the available P status of soils treated with two PRs varying in reactivity. The method was also able to describe the relationship between crop yield and available P for both PR and TSP, whereas the Olsen test failed. However, more work is needed to evaluate the mixed resin membrane test for various crops grown on different soils treated with PR and water-soluble P fertilizers, especially under field conditions.

Compared with the P_i strip test, the mixed cation-anion resin membrane strip has the advantage that the strip can be regenerated for reuse. However, it is only available from commercial chemical companies, and it is probably considerably more expensive than the P_i paper strip, especially for laboratories in developing countries.

Isotopic ^{32}P exchange kinetic method

The isotopic ^{32}P technique provides quantitative data on P behaviour in the soil (Fardeau, 1996) and soil-plant system for evaluating the efficiency of P fertilizers (e.g. PR products) as influenced by management practices (Zapata and Axmann, 1995). The isotopic ^{32}P exchange kinetic method (IEK) provides a full characterization of soil-available P by determining four independent factors in a single experiment. The intensity factor is quantified by the measurement of the chemical potential of P ions in the soil solution (C_p). The quantity factor is estimated by the quantity of P ions being instantaneously isotopically exchangeable at 1 min (E_1). Two capacity factors, one 'immediate' and the other 'delayed', are determined simultaneously. These four characteristics permit the application of a functional and dynamic model for soil-available P (Fardeau, 1996).

Most extraction methods used in soil testing only allow the determination of a quantity factor, whereas water extraction measures only the intensity factor. Although the IEK method allows quantification of intensity, quantity and capacity factors, it cannot be used in routine soil analysis because of the radioactive nature of the ^{32}P isotope. However, the IEK method can be considered as the reference method for assessing the suitability of other soil tests in routine determination of soil-available P. Aigner *et al.* (2002) compared available P measured by the P_i test with the IEK method in soils varying widely in soil properties treated with and without Morocco PR. For soils that did not receive PR, a very high correlation ($R^2 = 0.94$) was found between the P extracted by the P_i test and the E_1 value determined by the IEK method. For soils treated with PR, the correlation was lower ($R^2 = 0.53$). One of the soils with the highest C_p value (from Hungary) showed that the E_1 value decreased from 9.4 mg of P per kilogram without PR to 6.8 mg of P per kilogram with PR addition, whereas P_i-P value increased from 19.3 to 31.1 mg of P per kilogram. No definitive explanation was given for this discrepancy. After excluding the soil from Hungary, the correlation between the P_i test and IEK method was very high ($R^2 = 0.92$). Therefore, it was concluded that the P_i test is a good alternative to the isotopic method in routine analysis, providing better estimates of soil-available P than other extraction methods.

Chapter 7
Decision-support systems for the use of phosphate rocks

Phosphate rocks (PRs) can be used as phosphorus (P) fertilizer sources in agricultural systems. Under certain conditions, farmers can apply them in order to supply P to crops at a lower cost than water-soluble P fertilizer. However, before applying PR, a farmer will probably ask questions such as: "Will PRs work on my land?"; and "Would a water-soluble P fertilizer be more cost-effective?" Researchers who have worked with PRs will respond by saying: "It will depend on the type of PR that you use and the environmental conditions that occur in your fields". In fact, many factors determine whether a given PR will be an effective P fertilizer in a farmer's field. The way these factors interact to influence PR performance is complex for any specific condition. Thus, it is difficult to make general technical recommendations.

Decision-support systems (DSSs) are simple tools that enable research and extension personnel to provide technical recommendations and decision support to farmers. A DSS for PR use (PR-DSS) utilizes available information on the above factors to predict whether a given PR will be effective in a given crop environment. This chapter discusses different types of DSSs for predicting PR performance, including the approaches used to develop a PR-DSS in New Zealand and Australia. It presents examples that illustrate these approaches. Finally, it outlines steps that are being undertaken in the development a more global DSS by FAO, the International Fertilizer Development Center (IFDC) and the International Atomic Energy Agency (IAEA) for use in tropical and subtropical countries for a range of food crops.

THE NEED FOR A DECISION-SUPPORT SYSTEM FOR PHOSPHATE ROCKS

Farmers need to know whether the use of PR will: (i) be worthwhile in supplying P to a deficient soil; and (ii) will save money compared with using water-soluble P fertilizer. They could obtain this information from an expert experienced in PR use. However, there are limitations on accessing the information on these technical recommendations. This chapter contends that a DSS is the most effective way of integrating the main factors that determine PR effectiveness, and then of indicating whether the PR will be effective in farmers' fields.

A DSS is an interactive computer-based system that helps decision-makers utilize data and models to solve unstructured problems (Sprague and Carlson, 1982). The main aim of such a system is to improve the performance of decision-makers while reducing the time and human resources required for analysing complex situations. A PR-DSS will be able to predict whether a particular PR will be effective in providing P to the crop in the field. A number of these DSSs have been developed to predict PR performance in different environments.

CONCEPTUAL BASIS FOR BUILDING A PR-DSS

In practice, a PR will be agronomically effective in a farming system if it is able to dissolve fast enough to provide plant-available P at a rate that is adequate for crop growth. Thus, it is a question of whether the rates of PR dissolution can match the rates of P demand by crops. Various authors have reviewed the conditions necessary for this to occur in temperate (Khasawneh and Doll, 1978) and tropical regions (Hammond *et al.*, 1986b; Sale and Mokwunye, 1993). Chapter 5 discusses this aspect in more detail.

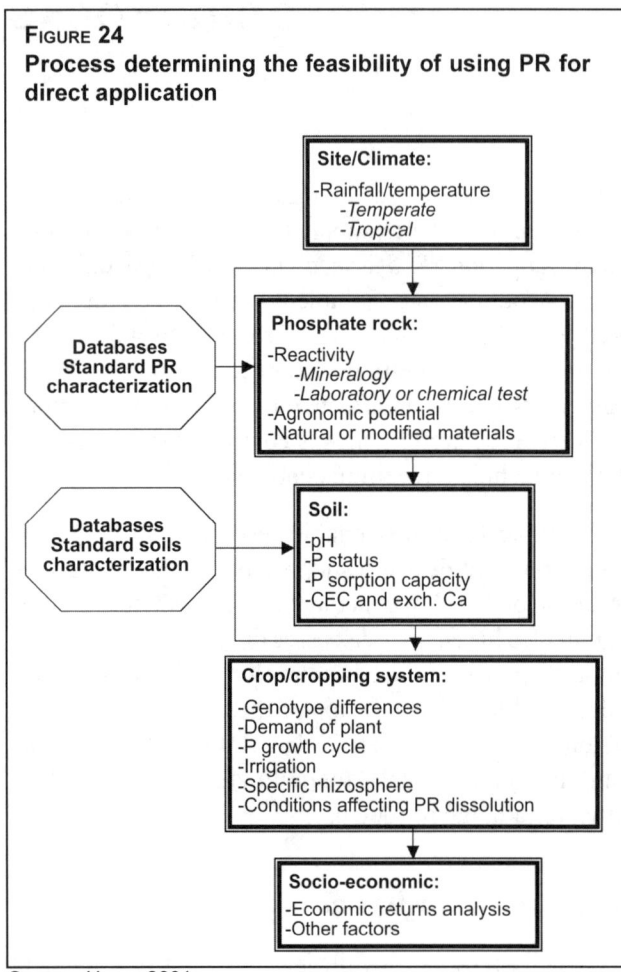

FIGURE 24
Process determining the feasibility of using PR for direct application

Source: Heng, 2001.

The conceptual framework in Figure 24 summarizes the factors that determine the feasibility of using PR for direct application (Heng, 2003). Five key factors determine whether the rate of supply of dissolved P from PRs will match the rate of P demand by the crop. These are: the reactivity of the PR, the properties of the soil, the prevailing climate conditions, the crop type, and the management system employed in growing the crop and applying the PR.

A final issue is the integration effect of time. A common feature of PRs is that the initial plant response to PR may be limited. However, their performance relative to water-soluble fertilizer tends to improve with time. This is because of the continuing dissolution of the PR compared with the declining availability of the P from residues and reaction products from water-soluble P sources. Thus, the PR can have a superior 'residual effect' compared with the water-soluble fertilizer. The farmer sees increasing benefits from the PR as time passes.

The framework in Figure 24 highlights the complexity of the whole system. Apart from the factors that affect the agronomic performance of the PR, many other issues will impinge on the use of PR. These include: the size of the potential PR deposit that would supply the PR; the cost of mining, grinding and distribution; the cost-benefit ratio for all participants in the supply chain; the social, economic and environmental impact; and government policy. In the final analysis, these will determine whether a deposit will be developed in order to enable farmers to use the PR as a fertilizer.

Chapter 10 discusses the issue of whether local farmers will adopt PR for use as a P fertilizer. The DSS instruments described in this chapter do not incorporate the question of farmer adoption and the socio-economic and policy issues outlined above. These DSSs focus solely on the question of whether the PR fertilizer will be effective in a specific agricultural system.

DIFFERENT TYPES OF DSS TO PREDICT PR PERFORMANCE

Mechanistic models

Perhaps the most complete approach for predicting how a PR will dissolve is the mechanistic model constructed by Kirk and Nye (1985a, 1985b, 1986). This model starts from the premise that the diffusion of the dissolving ions away from the PR surface is the rate-limiting step. The model is able to provide quite accurate predictions of PR dissolution under controlled conditions (Anderson and Sale, 1993). However, it cannot determine whether a particular crop will respond to a PR application in the field.

The major limitation to the use of complex mechanistic models is the need for a large number of input parameters that are relatively difficult to determine. For example, the Kirk and Nye model requires ten parameters in order to define the soil environment. These include: the concentration and activities of phosphate and calcium (Ca) ions in the soil solution, soil pH, pH buffering capacity, P buffering capacity, coefficients for P adsorption isotherms, soil bulk density, volumetric water content, and percentage clay. This approach is too complex for farm-level use of a PR-DSS.

The combination of mechanistic and empirical models

The New Zealand DSS

Researchers in New Zealand have developed a DSS for using PRs on high-rainfall areas with permanent pastures grazed by sheep and/or cattle. This system is based around five components (Perrott, 2003). The first is an empirical model that predicts the suitability of a particular pastoral site for PR use, given by the mean diffusion (Dm) coefficient of phosphate in soil at a specific site. This is based on direct dissolution measurements of a standard PR (0.075–0.150 mm Sechura PR) at 90 pastoral sites in New Zealand over a two-year period, which provided regression data to predict the Dm variable. Soil type, pH, exchangeable magnesium (Mg), soil drainage and rainfall are used as parameters to predict this variable.

The second component is a laboratory test that determines the reactivity of the PR by measuring the equilibrium P concentration sustained by the PR (CR) in a simulated soil solution maintained at a constant pH (5.5) using an automatic titrator. The CR value, together with the particle density, the P concentration, and the range and amount of particle sizes of the PR, provides a measure of the 'intrinsic reactivity of the PR' (Perrott, 2003).

The third component is a mechanistic model developed from a simplified version of the Kirk and Nye model (Watkinson, 1994a). It calculates the annual dissolution rate of the particular PR at the particular pastoral site, using the predetermined PR and site variables. This annual dissolution rate is then expressed as a rate constant for use in an exponential dissolution model (fourth component) and the annual release of P from the PR is then treated as a plant-available P input into the labile soil P pool, for the Overseer™, which is an integrated fertilizer DSS (fifth component). Overseer™ is a tool for making site-specific recommendations on the use of P, potassium (K) and sulphur (S) fertilizers on the basis of economic and environmental criteria (Metherell and Perrott, 2003).

The New Zealand DSS is specific for reactive PRs on grazed pasture systems on New Zealand soil types. While the system is limited to this form of land use in this part of the world, the approach has much to recommend it. It attempts to define and quantify the sequential steps in the PR dissolution process. By determining empirical predictors based on simple field experiments, it simplifies complex soil-site-climate interactions that determine the likelihood of PRs dissolving rapidly in a particular pastoral environment. By determining the amount of dissolved PR entering the plant labile pool each year, the modelling process enables the PR to be compared with water-soluble P fertilizers, which also contribute plant-available P to the labile pool in the soil.

The New Zealand DSS is the product of many years of research by chemists, agronomists and computer specialists. The importance of the grazing industries to New Zealand and the need for nutrient inputs to maintain the productivity of these industries have sustained this research effort.

The PR-DSS being developed by the IFDC

The IFDC has also developed a preliminary version of a PR-DSS for estimating the agronomic effectiveness of freshly applied PR with respect to water-soluble phosphate (WSP) fertilizers (Hellums *et al.*, 1992; Chien *et al.*, 1999; Singh *et al.*, 2003). The development is based on IFDC work in West Africa. The model incorporates the effect of PR sources (PR solubility), soil pH, soil texture, organic matter, crop type, and moisture/rainfall regime in predicting the relative agronomic effectiveness (RAE) of the PR with respect to WSP fertilizers. The current version of the model is not able to determine whether P is limiting or what the P fertilizer rate should be. It assumes that other nutrients and pests are non-limiting. It does not consider any socio-economic evaluation.

An example of the use of the current IFDC PR-DSS

A task for the IFDC PR-DSS was to predict how effective Minjingu PR from the United Republic of Tanzania might be for growing a maize crop on a soil at Kabete, Kenya (Singh *et al.*, 2003). The DSS required the following inputs: the neutral ammonium citrate (NAC) solubility of the moderately reactive PR (8.45 percent P_2O_5 of the PR), the soil pH_{water} (5.02), the organic carbon percent of the soil (1.44 percent), the clay content (25 percent), the sand content (38 percent), the growing-season rainfall (800 mm), and the crop type (maize). The DSS predicted the RAE for Minjingu PR would be 85 percent. Singh *et al.* (2003) reported that the observed RAE values at this site ranged from 68 to 100 percent, with the value for the fourth annual application ranging from 80 to 90 percent.

EXPERT SYSTEMS

It is also possible to use expert systems to develop a DSS. They do not require the years of dedicated research needed to develop mechanistic models. Instead, they require the input of a skilled 'knowledge engineer', who spends substantial time working with PR specialists, and other experts who have years of experience in the field. Australian researchers used this approach in order to construct a DSS to advise on PR use on pastures. The task was to determine where PRs might be effective on pastures grazed by sheep and/or cattle in high-rainfall regions in eastern and southern Australia.

The 'RPR Adviser' in Australia

An expert system was built on the findings of a large, national project that managed to establish 25 effective, replicated field experiments on representative soils in the target regions. The project generated annual substitution values of triple superphosphate (TSP) for the highly reactive North Carolina PR at each site for a four-year period. The SV50 parameter is the amount of P (kilograms) supplied as TSP that produced 50 percent of the maximum yield response to P at a particular site, divided by the amount of P (kilograms) from North Carolina PR required to produce the same yield.

The project provided a picture of the performance of PRs in a range of pastoral environments that differed in rainfall, soil properties, and pasture types. These involved environments where North Carolina PR was: (i) as effective as TSP in the first year; (ii) almost as effective as TSP over time; and (iii) completely ineffective compared with TSP. The effect of time was allowed for as PR effectiveness was considered after 1 and 4 years of annual PR applications. A 'knowledge engineer' then used this empirical information, together with scientific understanding of the way the four environmental variables determine PR effectiveness, to construct an expert system called 'RPR Adviser' that could advise Australian pastoral farmers on PR use (Gillard *et al.*, 1997).

The approach taken in building the RPR Adviser was to assign a scalar weighting to each climate or soil factor that appeared to affect PR performance. These included: the reactivity of the PR, rainfall, soil pH, soil texture, the likelihood of leaching of P from water soluble fertilizer, P sorption capacity, and pasture composition. Each factor was assigned independently, with the overall prediction being achieved by multiplying all the factor weightings together.

The expert system was able to provide accurate predictions of the SV50 values that occurred (the correlation coefficient between predicted and observed SV50 values was 0.92) (Gillard *et al.*, 1997). It was also possible to validate results from a number of independent field experiments by using pasture yield responses to PR and superphosphate fertilizer. The required input variables for the expert system were simple and did not require specialized laboratory analyses. The site variables were: annual rainfall, the likelihood that soluble P will leach from the rootzone, and pasture composition. The required soil variables were: pH, Colwell P, field texture and soil colour, with colour and texture being surrogates to estimate P sorption capacity for the soil.

The researchers found some useful hints for using the computerized DSS. As there was no commercial intellectual property involved, the RPR Adviser has been made available free-of-charge at: www.latrobe.edu.au/www/rpr/. This has facilitated its widespread use. In addition, by having notes and string messages, it is possible to explain how a decision has been derived. Good 'help' facilities can help the user to use the system with confidence.

Farmers in Australia received a more simplified DSS in the form of a checklist of questions. The questions gave an indication of conditions likely to result in effective PR use on Australian pastures. If all the questions could be answered in the affirmative, then there was a high likelihood that reactive PRs would be useful P fertilizers at the specific pasture site.

An example of the use of the RPR Adviser

A beef producer in southern Gippsland, a high-rainfall district in southeast Australia, wanted to determine whether a reactive PR might be an effective P fertilizer, compared with single superphosphate, for maintaining the productivity of white-clover, perennial ryegrass pastures. The farmer downloaded the RPR Adviser from the Internet and answered the questions asked by the DSS. The answers were that the annual rainfall was 1 000 mm, the PR to be used was highly reactive, the soil was not red or lateritic (indicating that the P sorption capacity was not high), the soil had a deep sandy texture, and the soil pH_{water} was 5.3. Unsure of what was meant by a question about the extent of leaching in the soil, the farmer checked the help file and received a brief explanation. The farmer then gave the answer that moderate leaching was possible. The DSS responded by indicating that the highly reactive PR would be as effective as superphosphate in the first year if the farmer changed to the RPR. This prediction is consistent with RPR performance in similar environments across southern Australia. The DSS then advised the beef producer to make an economic assessment of the choice based on the cost per kilogram of P applied. Furthermore, the DSS also warned of a possible shortage of S in that pasture soil if straight RPR were used without added S.

REQUIREMENTS FOR A GLOBAL DSS FOR PR USE

Heng (2003) has outlined a conceptual framework (Figure 24) for the further development of a more global PR-DSS. It is possible to prescribe a number of steps and a series of guidelines for the development of this kind of system. Such a system is now required in order to provide advice on PR effectiveness for food and cash crops in the tropics and subtropics, particularly in developing countries.

Creation of a database for PR reactivities

Phosphate-bearing minerals vary widely in their inherent characteristics. Therefore, the characterization of the agronomic potential of PRs is the first and essential step in evaluating their suitability for direct application and for the development of a comprehensive DSS to evaluate PR use. The specific tasks include: (i) collate data on PR solubility analysed with NAC (second extraction) with standard procedures (specified shaker speed and time); (ii) correlate the solubility data obtained with other available data for reactivity tests of the same PRs; (iii) conduct a regression analysis to establish relationship between the two methods under study; (iv) analyse outliers using standard procedures; and (v) analyse all remaining samples using standard procedures.

The Joint FAO/IAEA Division of Nuclear Techniques in Food and Agriculture and the IFDC will be sharing their data set of PR properties in order to create a global database for a PR-DSS. Over the years, the IFDC has conducted numerous field trials in Africa, Asia and Latin America and it has accumulated a large database. Similarly, through the international PR research network implemented from 1993-99 in both developing and developed countries, the Joint FAO/IAEA Division of Nuclear Techniques in Food and Agriculture has collected valuable data on the agronomic effectiveness of P fertilizers, with a variety of PR sources and numerous field crops in a wide range of environments (Zapata, 2000, 2003; IAEA, 2002).

Creating a database of soils for PR use

A standard characterization of soil samples from target regions where PRs might be considered for use is required in order to provide reliable input information for the DSS. This will involve collating measurements for key soil properties that affect the dissolution of PRs. The standard characterization of both PRs and soil properties was undertaken by the Joint FAO/IAEA Division of Nuclear Techniques in Food and Agriculture from the research network project on "The use of nuclear and related techniques for evaluating the agronomic effectiveness of phosphatic fertilizers, in particular rock phosphates" (Truong and Zapata, 2002; Montange and Zapata, 2002).

Other requirements

The PR performance information that is used to construct a DSS needs to cover the full range of PRs, climates, soil types and farming systems that are likely to be encountered in the target region. For example, if a DSS is required for certain regions in Africa, then all of the possible findings on PR performance in those regions need to be incorporated in the DSS. This will include socio-economic factors that are an integral part of the farming system. These may well determine whether a local PR might be used in a specific cropping situation. There is an emerging view that DSS applications need to focus on real problems facing ordinary people, and attempt to develop solutions to them (Matthews *et al.*, 2002). The provision of cost-effective plant nutrient sources to combat nutrient-depleted soils is a complex problem throughout the developing world.

The input information for the DSS needs to be simple and able to be provided by likely users. It is not practical to request complex soil measurements, such as pH buffering capacity, as an input variable for the DSS if the user has no idea of the measurement and it can only be obtained in a research laboratory. Soil colour, texture, pH and rainfall are examples of readily obtainable input variables.

There also needs to be close collaboration between the people building the DSS and the researchers who are evaluating PRs in the field. The former need to receive all possible information about PR performance and the environment in which the PR was used, while the latter need to provide all possible information on PR performance for incorporation in the DSS. Researchers planning a field evaluation of a PR should consult with the DSS developers in order to determine whether there might be an additional treatment to include in the field experiment that could be useful in developing the DSS further.

FUTURE DEVELOPMENTS

Both the IFDC and FAO/IAEA have agreed to collaborate in the development of a global PR-DSS. Plans for this development are underway in terms of developing a simplified PR database from the large amount of information available. Further development will require the processing of additional data in order to establish functional and statistical relationships and then their inclusion in the PR-DSS.

A network of pilot testing and validation field experiments will also be established in a wide range of agro-ecological zones. These will be located at selected sites in Africa, Asia and Latin America and gather information at the cropping-system level, including agronomic management and socio-economic data. This will enable enhanced prediction of expected results and improved decision-making. The resulting DSS will be further developed as a Web-based system, delivering outputs over the Internet. This will facilitate resource sharing, adherence to corporate standards and use of tools and standards developed by the World Agricultural Information Center.

The availability of a global PR-DSS will be a useful research and extension tool for researchers, extension workers, farmers, planners, and agribusiness dealers. It will assist in promoting the use of PR resources in tropical and subtropical developing countries.

Chapter 8
Secondary nutrients, micronutrients, liming effect and hazardous elements associated with phosphate rock use

Phosphate rock (PR) is recommended for application to acid soils where phosphorus (P) is an important limiting nutrient on plant growth. The past 50 years have seen the accumulation of considerable knowledge regarding the factors affecting the agronomic effectiveness of PR for direct application. However, much less information is available on other effects associated with PR use, i.e. secondary nutrients, micronutrients, liming effect, and hazardous elements. This chapter presents a review of the information available in literature that is relevant to these other effects.

SECONDARY NUTRIENTS IN PHOSPHATE ROCK

Among the significant chemical and nutritional constraints on crop growth on acid soils are deficiencies of calcium (Ca) and magnesium (Mg) nutrients. As the apatite mineral in PR is Ca-P, there is a potential to provide Ca nutrient if there are favourable conditions for apatite dissolution. Furthermore, many sources of PR contain free carbonates, such as calcite ($CaCO_3$) and dolomite ($CaMg(CO_3)_2$), that can also provide Ca and Mg in acid soils. However, if dissolution of free carbonates raises pH and exchangeable Ca around PR particles significantly, it can hinder apatite dissolution and thus reduce P availability of PR (Chien and Menon, 1995b). For example, Chien (1977) found that Huila PR (Colombia), which contained about 10 percent $CaCO_3$, increased soil solution pH from 4.8 to 6.2 in one week compared with other PRs that increased pH to 5.1. Consequently, the maximum soil solution P concentration obtained with Huila PR was lower than that obtained with central Florida PR (Figure 25), even though the two PR sources had approximately the same degree of isomorphic substitution of CO_3 for PO_4 in apatite structure.

Hellums *et al.* (1989) reported on the potential agronomic value of Ca in some PRs from South America and West Africa. Their study applied adequate P as KH_2PO_4 to an acid sandy loam (pH 4.5) with low exchangeable Ca to isolate the Ca from the P effect. The results showed that Ca uptake by maize with various PR sources followed the order of the reactivity of the PRs except Capinota PR (Bolivia), which had about 10 percent $CaCO_3$ (Figure 26). The relative agronomic effectiveness (RAE) of various PR sources with respect to $CaCO_3$ (100 percent) in terms of increasing dry-matter yield and Ca uptake ranged from 28 to 89 percent and from 8 to 58 percent, respectively (Table 26). The results showed that PRs of medium and high reactivity have potential Ca value, in addition to their use as a P source, when applied directly to acid soils with low exchangeable Ca.

In a three-year field trial conducted in central China, Hu *et al.* (1997) reported that exchangeable Ca increased from 1 194 mg/kg with the control to 1 300–2 100 mg/kg with

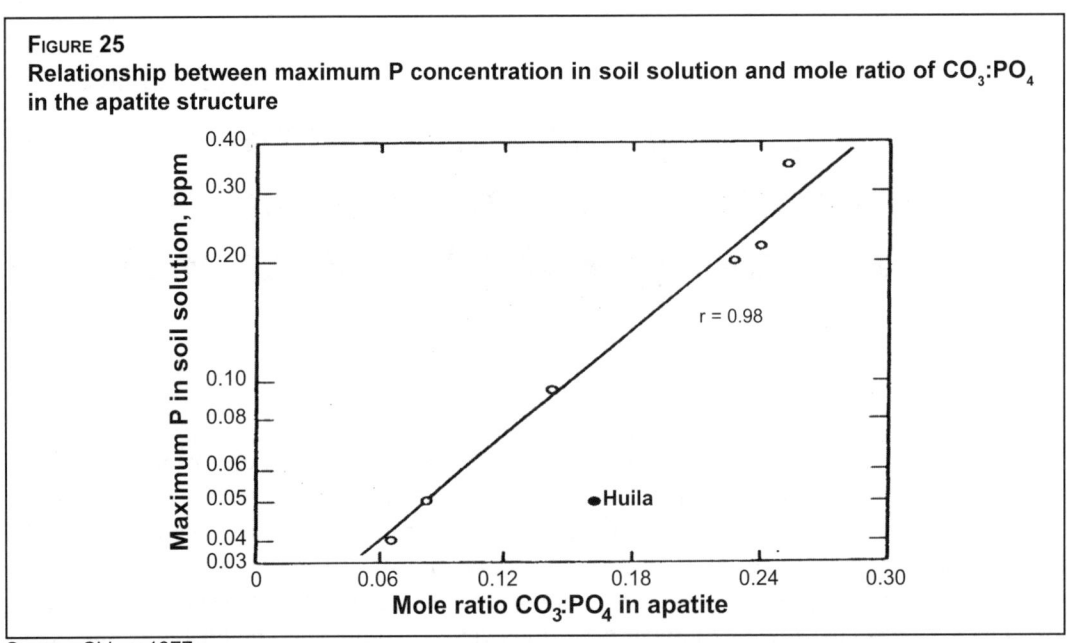

FIGURE 25
Relationship between maximum P concentration in soil solution and mole ratio of $CO_3:PO_4$ in the apatite structure

Source: Chien, 1977a.

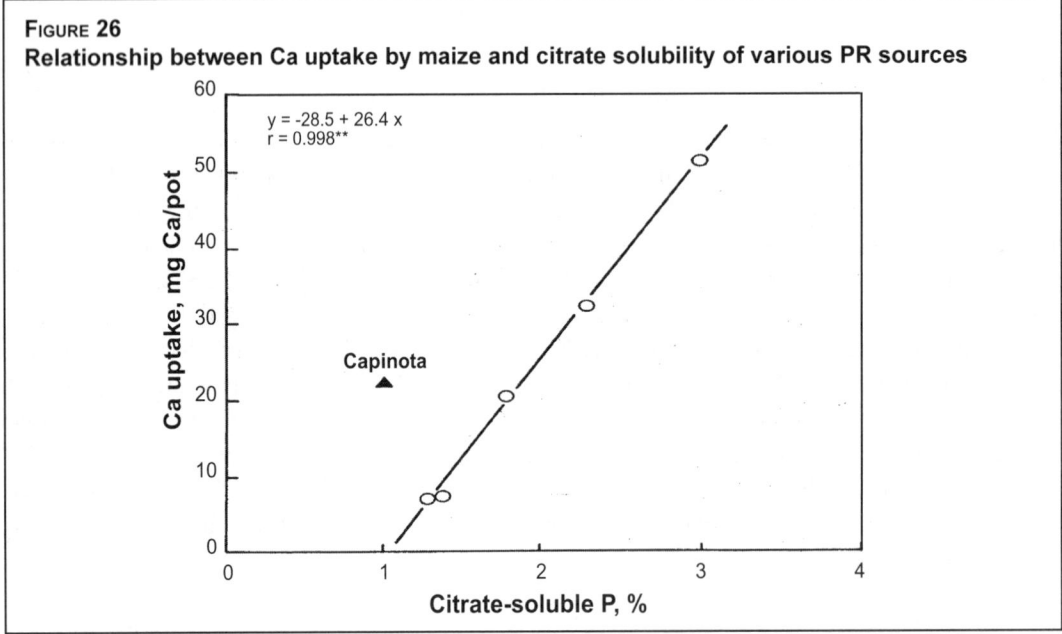

FIGURE 26
Relationship between Ca uptake by maize and citrate solubility of various PR sources

Source: Hellums et al., 1989.

TABLE 26
Relative agronomic effectiveness of various PRs with respect to $CaCO_3$ as a Ca source for maize

Ca source	Reactivity	Relative agronomic effectiveness (%)	
		Dry-matter yield	Ca uptake
Bahia Inglesa PR (Chile)	High	89	58
Bayovar PR (Peru)	High	73	33
Capinota PR (Bolivia)	Low	52	17
Tilemsi Valley PR (Mali)	Medium	53	17
Tahoua PR (Niger)	Low	31	8
Hahotoe PR (Togo)	Low	28	8
$CaCO_3$		100	100

Source: Hellums et al., 1989.

PR treatments. The corresponding exchangeable Mg levels were 330 mg/kg with the control and 350–400 mg/kg with the PR treatments. Because the content of apatite-bound Mg is very small (unlike apatite-bound Ca), it is expected that PR will not increase soil exchangeable Mg significantly unless the PR contains a significant amount of dolomite. More research is needed to obtain information on the agronomic value of Ca and Mg (especially the latter).

Some PR sources may contain a significant amount of sulphur (S) bearing accessory minerals, e.g. gypsum ($CaSO_4$) in Israeli PR (Axelrod and Gredinger, 1979) and pyrite (FeS_2) and pyrrhotite (FeS) in Mussoorie PR, India (PPCL, 1983). However, little information is available on S availability to plants from these PR sources.

MICRONUTRIENTS IN PHOSPHATE ROCK

Some PRs contain accessory minerals that may provide micronutrients to aid plant growth. However, limited information is available on this potential additional benefit of using PR for direct application.

Work by Hammond *et al.* (1986b) on an Oxisol in Colombia suggested that indigenous Huila PR, which contains 136 mg of zinc (Zn) per kilogram, produced a higher grain yield of one rice variety (Cica-8) than did triple superphosphate (TSP) because of its Zn content (Figure 27). However, available Zn from Huila PR alone was not sufficient to provide adequate Zn for two rice varieties. When Zn was applied to the soil, both Huila PR and TSP were equally effective in increasing rice grain yield.

In New Zealand, Sinclair *et al.* (1990) found that Sechura PR (Peru), which contains 43 mg of molybdenum (Mo) per kilogram, increased dry-matter yields of pasture herbage more than TSP did at sites where the PR increased Mo levels in clover significantly (Figure 28). More information is needed on the micronutrient contents of PRs that have potential for increasing crop production on acid soils.

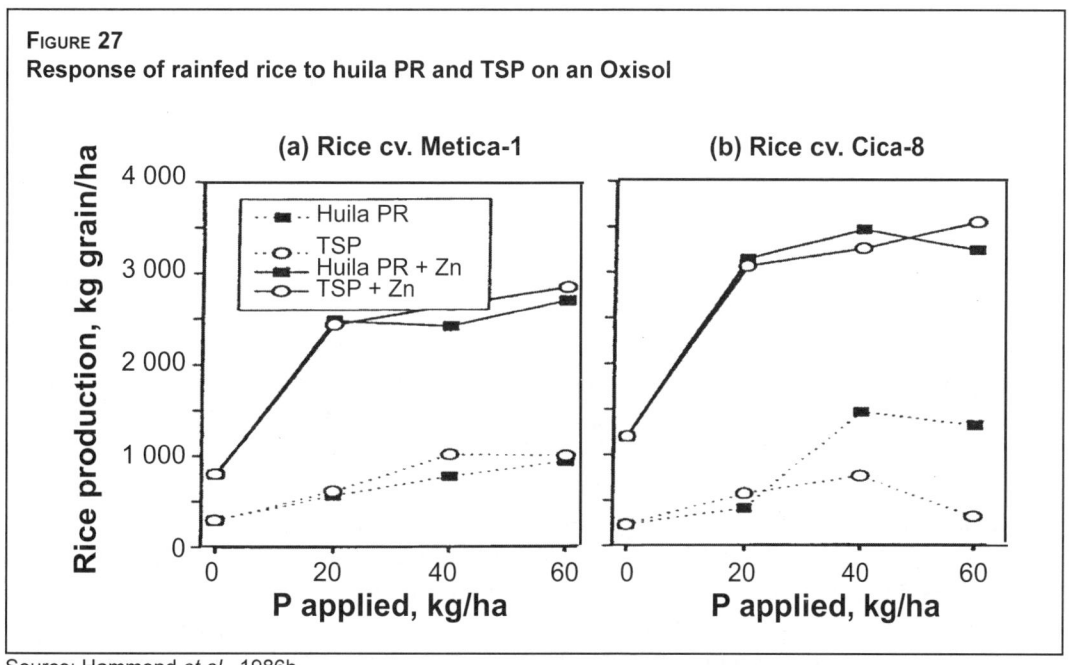

FIGURE 27
Response of rainfed rice to huila PR and TSP on an Oxisol

Source: Hammond *et al.*, 1986b.

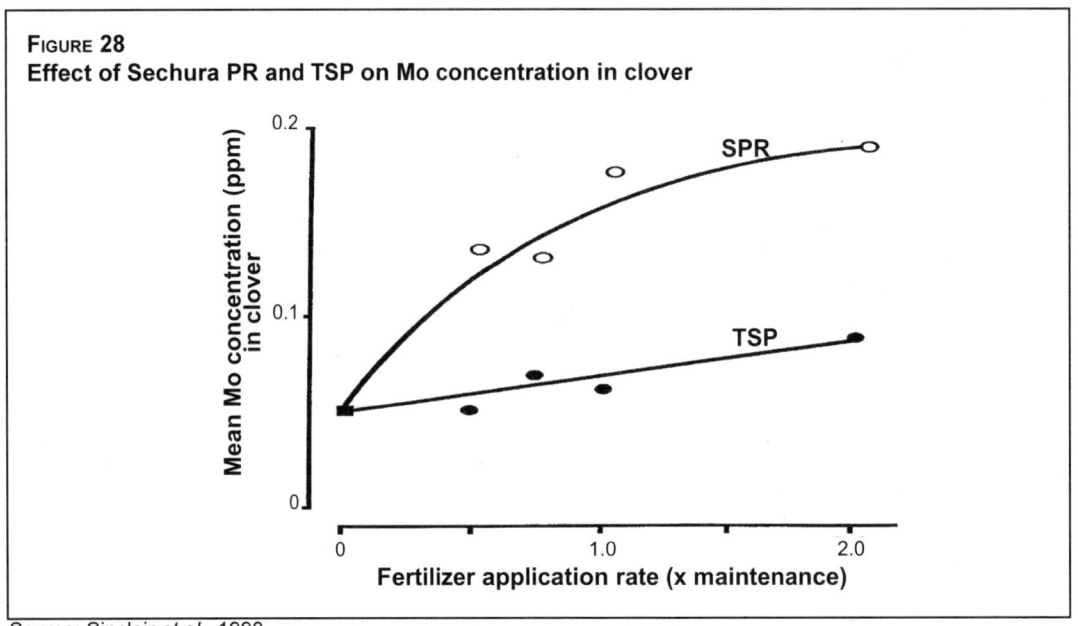

FIGURE 28
Effect of Sechura PR and TSP on Mo concentration in clover

Source: Sinclair et al., 1990.

LIMING EFFECT ASSOCIATED WITH PR USE

Low pH coupled with toxic levels of aluminium (Al) and manganese (Mn) frequently contributes to poor soil fertility for plant growth on acid tropical and subtropical soils in developing countries. Although lime is effective in alleviating soil acidity and Al toxicity, it is often either unavailable or expensive to transport. Screening crop species and varieties to identify those that are tolerant of soil acidity would reduce lime requirements (Sanchez and Salinas, 1981; Goedert, 1983).

The dissolution of apatite in PR consumes H^+ ions and, thus, it can increase soil pH, depending on PR reactivity. If a PR contains a significant amount of free carbonates, it can further increase soil pH. However, although an increase in soil pH may reduce the Al saturation level, it can also reduce apatite dissolution at the same time. The optimum condition would call for a soil pH that is high enough to reduce the Al saturation level but still low enough for apatite dissolution to release P.

Research by the International Fertilizer Development Center (IFDC) has shown that the application of medium to highly reactive PRs with low free-carbonate contents can result in significant liming effects on acid soils. Although the increase in pH is generally less than 0.5 units, the decrease in exchangeable Al can be significant where the soil pH is less than 5.5 (Chien and Friesen, 2000) as the exchangeable Al level would be almost zero at this soil pH in Oxisols and Ultisols (Pearson, 1975). For example, exchangeable Al was reduced from 2.0 to 0.4 meq/100 g when a Colombian Oxisol was treated with Sechura PR in a soil incubation study (Figure 29). The soil pH increased correspondingly from 4.6 to 5.0, and the exchangeable Ca rose from 0.2 to 1.5 meq/100 g. Consequently, the Al saturation level also declined from about 80 to 20 percent. Thus, the better performance of highly reactive PRs, e.g. Sechura and North Carolina, compare with TSP in plant-growth response on the soil may have been related to the alleviation of Al toxicity (Figure 30). In a five-year field trial conducted in an Oxisol fertilized with various PR sources, Chien et al. (1987b) reported that the pH increased from 4.1 with the control to 4.7–5.0 with the PR treatments. The corresponding increase in exchangeable Ca was from 0.17 cmol/kg with the control to 0.31–0.56 cmol/kg with the PR treatments. However, no

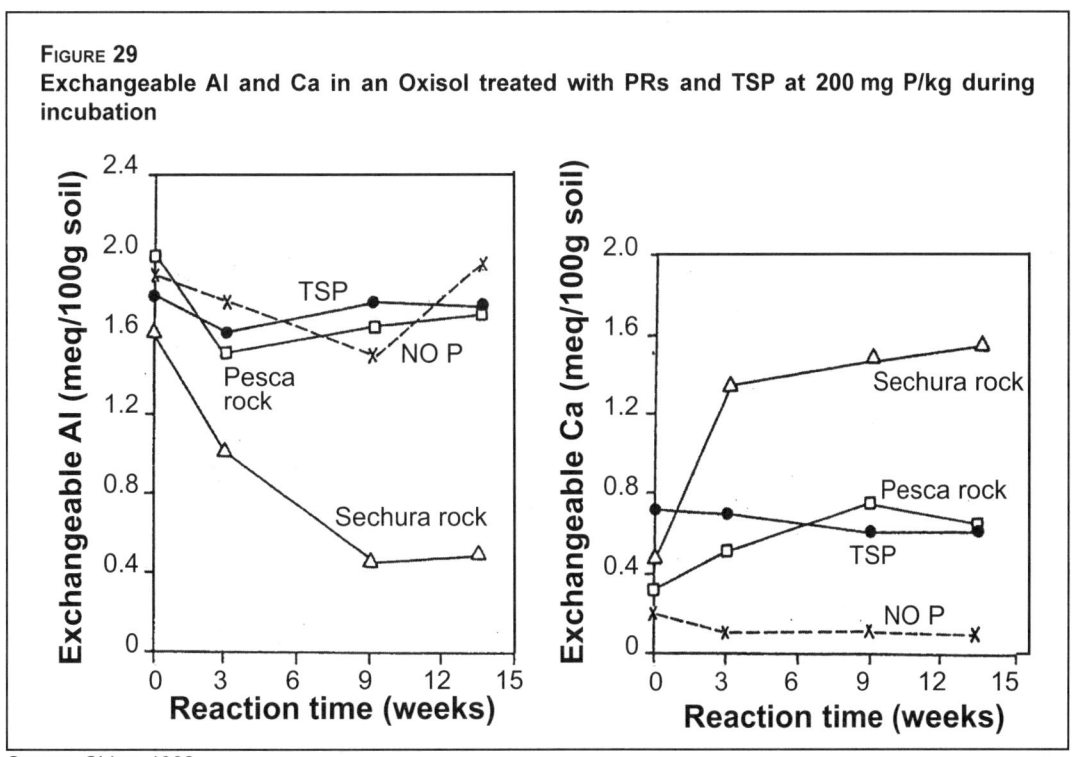

FIGURE 29
Exchangeable Al and Ca in an Oxisol treated with PRs and TSP at 200 mg P/kg during incubation

Source: Chien, 1982.

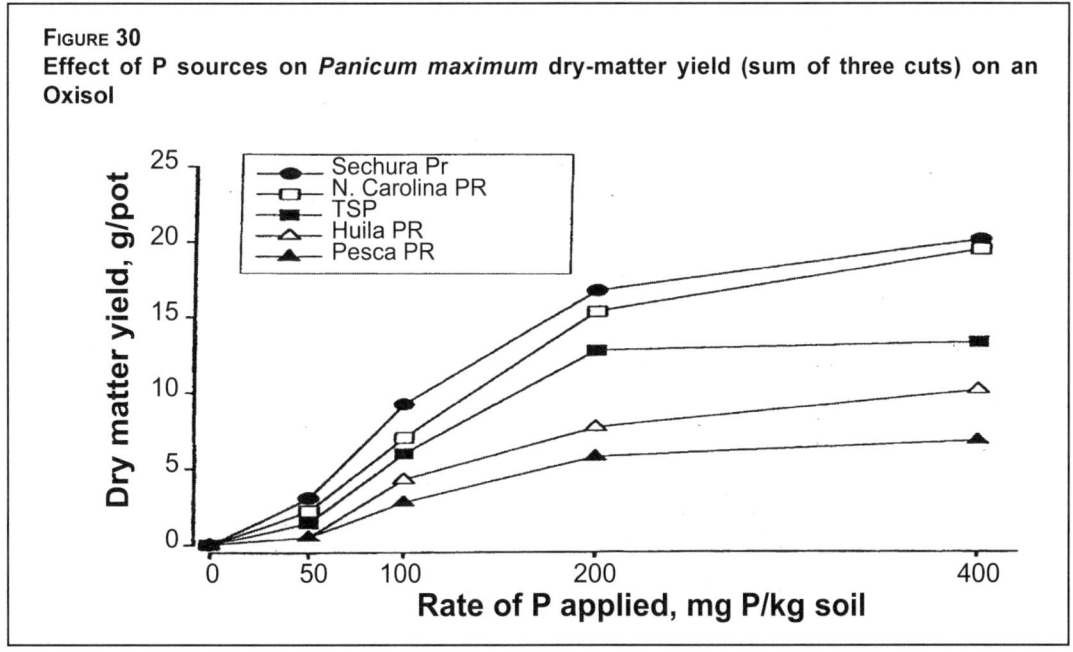

FIGURE 30
Effect of P sources on *Panicum maximum* dry-matter yield (sum of three cuts) on an Oxisol

Source: Chien, 1982.

significant effect on exchangeable Al was observed. In their study with the red soil of China, Hu *et al.* (1997) reported that the soil pH increased from 4.8 with the control to 4.9–5.3 with the PR treatments. A reduction in exchangeable Al of up to 70 percent with respect to the control was also observed with the PR treatments. Thus, the studies suggest that the application of PR to acid soils can also improve soil properties as well as the supply of available P for crop production.

Sikora (2002) conducted a theoretical and experimental study to calculate and quantify the liming potential of PRs by laboratory titration and soil incubation. Of the three anions (PO_4^{-3}, CO_3^{-2} and F^-) present in the carbonate apatite structure of PR, CO_3^{-2} and PO_4^{-3} can consume H^+ and cause an increase in pH. Because of the greater molar quantity of PO_4^{-3} compared with CO_3^{-2}, PO_4^{-3} exerts a greater effect on the liming potential of PR. The results for the titration of two PRs (highly reactive North Carolina and low-reactive Idaho) showed the ranges of calcium carbonate equivalence (CCE) were from 39.9 to 53.7 percent, which were less than the theoretical values (59.5 to 62.0 percent). The experimental model obtained from the soil incubation study showed qualitative agreement with theory as it showed increased liming ability with increased dissolved P from the PRs. However, the model showed lower percentage CCEs than theoretical calculations when the P dissolved ranged from 20 to 60 percent. Further research is needed to compare actual percentage CCE models across a variety of soil types in order to assess the potential liming effect associated with PR use.

HAZARDOUS ELEMENTS IN PHOSPHATE ROCK

All PRs contain hazardous elements including heavy metals, e.g. cadmium (Cd), chromium (Cr), mercury (Hg) and lead (Pb), and radioactive elements, e.g. uranium (U), that are considered to be toxic to human and animal health (Mortvedt and Sikora, 1992; Kpomblekou and Tabatabai, 1994b). The amounts of these hazardous elements vary widely among PR sources and even in the same deposit. Table 27 shows the results of a chemical analysis of potentially hazardous elements in some sedimentary PR samples (Van Kauwenbergh, 1997).

Among the hazardous heavy metals in P fertilizers, Cd is probably the most researched element. This is because of its potentially high toxicity to human health from consuming foods that are derived from crops fertilized with P fertilizers containing a significant amount of Cd. Most of the studies on Cd uptake by crops have used water-soluble P fertilizers such as TSP, single superphosphate (SSP), di-ammonium phosphate and mono-ammonium phosphate. However, the Cd reaction with soil treated with PR differs significantly from that with water-soluble P fertilizers because apatite-bound Cd in PR is water insoluble. Iretskaya et al. (1998) reported

TABLE 27
Chemical analysis of potentially hazardous elements in sedimentary phosphate rocks

Country	Deposit	Reactivity	P_2O_5 (%)	As (mg/kg)	Cd (mg/kg)	Cr (mg/kg)	Pb (mg/kg)	Se (mg/kg)	Hg (µg/kg)	U (mg/kg)	V (mg/kg)
Algeria	Djebel Onk	High	29.3	6	13	174	3	3	61	25	41
Burkina Faso	Kodjari	Low	25.4	6	<2	29	<2	2	90	84	63
China	Kaiyang	Low	35.9	9	<2	18	6	2	209	31	8
India	Mussoorie	Low	25.0	79	8	56	25	5	1 672	26	117
Jordan	El Hassa	Medium	31.7	5	4	127	2	3	48	54	81
Mali	Tilemsi	Medium	28.8	11	8	23	20	5	20	123	52
Morocco	Khouribga	Medium	33.4	13	3	188	2	4	566	82	106
Niger	Parc W	Low	33.5	4	<2	49	8	<2	99	65	6
Peru	Sechura	High	29.3	30	11	128	8	5	118	47	54
Senegal	Taiba	Low	36.9	4	87	140	2	5	270	64	237
Syrian Arab Republic	Khneifiss	Medium	31.9	4	3	105	3	5	28	75	140
United Republic of Tanzania	Minjingu	High	28.6	8	1	16	2	3	40	390	42
Togo	Hahotoe	Low	36.5	14	48	101	8	5	129	77	60
Tunisia	Gafsa	High	29.2	5	34	144	4	9	144	12	27
United States of America	Central Florida	Medium	31.0	6	6	37	9	3	371	59	63
United States of America	North Carolina	High	29.9	13	33	129	3	5	146	41	19
Venezuela	Riecito	Low	27.9	4	4	33	<2	2	60	51	32

Source: Van Kauwenbergh, 1997.

highly reactive North Carolina PR containing 47 mg of Cd per kilogram was as effective as SSP produced from the same PR in increasing grain yield of upland rice, but that the Cd concentration in rice grain with the PR was only about half of that with SSP. Thus, the information on Cd availability from water-soluble P sources cannot be implied directly to PR application.

The reactivity of the PR influences the availability of Cd to the plant because Cd is bound with P in the apatite structure (Sery and Greaves, 1996). To separate the P effect on Cd availability from PR, Iretskaya *et al.* (1998) pretreated two acid soils with 200 mg of P per kilogram as KH_2PO_4 so that no P response from PR would be expected in increasing grain yield of upland rice. They found that total Cd uptake by rice from the low-reactive Togo PR was 80 percent of that from the highly reactive North Carolina PR in the soil with a pH of 5.0, and 52 percent in the soil with a pH of 5.6 when the soils were treated with 400 μg of Cd per kilogram from the two PRs. McLaughlin *et al.* (1997) found that Cd concentrations in clover grown on the soil treated with a lower reactive Hamrawein PR (Egypt) containing 5.3 mg of Cd per kilogram were lower than that treated with highly reactive North Carolina PR containing 40.3 mg of Cd per kilogram at the same P rates. Thus, a PR source with a higher reactivity and Cd content can release more Cd than a PR with a lower reactivity and/or low Cd content for plant uptake. In addition to PR reactivity and Cd content, plant uptake of Cd also depends on soil properties, especially soil pH, and crop species (Iretskaya and Chien, 1999). More research is needed to investigate their interactions and integrate these factors on Cd availability associated with PR use.

Some PR sources may contain a significant amount of radioactive elements compared with other PR sources, e.g. 390 mg of U per kilogram in Minjingu PR (the United Republic of Tanzania) versus 12 mg of U per kilogram in Gafsa PR (Tunisia) (Table 27). As Minjingu PR is highly reactive and agronomically and economically suitable for direct application to acid soils for crop production (Jama *et al.*, 1997; Weil, 2000), there has been concern over the safety of using this PR. Samples of soil and plant tissue associated with the use of this PR were collected by the International Centre for Research in Agroforestry and sent to the International Atomic Energy Agency for radioactivity testing. The results showed that the radioactivity of the soil and plant samples was about the same as the background levels. However, the potential safety problem remains a concern for workers during mining operations.

Most PRs also have high concentrations of fluorine (F) in apatite minerals, often exceeding 3 percent by weight (250 g of F per kilogram of P). Excessive F absorption has been implicated in causing injury to grazing stock through fluorosis. McLaughlin *et al.* (1997) reported no significant differences between F in herbage from plots fertilized with either SSP containing 1.7 percent F or North Carolina PR containing 3.5 percent F, or between sites that had received both fertilizers. Concentrations of F in herbage were generally less than 10 mg of F per kilogram and often near the detection limit for the analysis technique (1 mg of F per kilogram). They concluded that plant uptake of F is unlikely to lead to problems for grazing animals in most soils. However, they cautioned that ingestion of soil by animals or ingestion of fertilizer material remaining on herbage after heavy topdressing could affect animal health, depending on soil and fertilizer F concentrations. Thus, there is a need to manage F in PRs in long-term applications to acid soils.

Chapter 9
Ways of improving the agronomic effectiveness of phosphate rocks

Chapters 5 and 7 show that the agronomic performance of phosphate rocks (PRs) applied directly as phosphate fertilizers depends on various factors and their interactions. The main factors include: (i) the physical and chemical properties of PRs; (ii) soil and climate factors; (iii) plant species and the cropping system; and (iv) farming management practices. There are situations where directly applied PRs are not effective. In these cases, it is possible to use various means to increase their agronomic effectiveness and thereby make the products more economically attractive. This chapter discusses the various possibilities under the headings of biological, chemical and physical means.

BIOLOGICAL MEANS

The biological means of enhancing the agronomic effectiveness of PRs applied as phosphate fertilizers are: (i) composting organic wastes with PR (phospho-composts); (ii) inoculation of seeds or seedlings with phosphorus-solubilizing micro-organisms (fungi, bacteria and actinomycetes); and (iii) the inclusion in the cropping system of crop genotypes that exhibit greater root growth and thus increase the extent of soil exploration, exude proton and organic acids that increase the solubility of sparingly soluble phosphates by decreasing pH and/or chelation, and produce elevated levels of phosphatase enzymes that can break organic phosphorus (P) down to inorganic P.

Phospho-composts

Treating PRs with organic materials and composting them is a promising technique for enhancing the solubility and the subsequent availability to plants of phosphorus (P) from PRs. The technology is particularly attractive where: (i) 'moderate to high' reactive PRs are available but unsuitable for the production of fully acidulated fertilizers such as single or triple superphosphate; (ii) organic manures are applied routinely to maintain the organic fraction of soils and supplement their nutrient requirement (as in most tropical countries); (iii) organic farming is practised, which excludes the use of chemically processed fertilizers; and (iv) city and farm by-products need to be disposed of in an environmentally friendly manner. The PR-composted products are usually referred to as phospho-composts.

Principles of phospho-composting

Phospho-composting is based on sound scientific principles. During the decomposition of organic materials, intense microbial activity occurs. This results in numerous types of bacteria and fungi that produce a large number of organic acids and humic substances. Some of the most commonly produced organic acids are: citric, malic, fumaric, succinic, pyruvic, tartaric, oxaloacetic, 2-ketogluconic, lacticoxalic, propionic and butyric (Stevenson, 1967).

The term humic substances is a generic name given to a large number of amorphous, colloidal organic polymers formed during the decomposition of organic matter. Humic substances are of high molecular weight and generally more stable than organic acids. Humic substances can be divided into three main fractions based on their solubilities in acid and/or alkali. The fraction that is soluble in acid and alkali is called fulvic acid, that soluble in alkali but precipitated in acid is humic acid, and that insoluble in acid and alkali is humin. Of these, the fulvic acid fraction has the lowest molecular weight, followed by humic acid and humin.

The enhancement of P release from PRs seems to be a function of the acidification of PR by organic acids and more importantly their chelating ability on calcium (Ca), iron (Fe) and aluminium (Al) (Pohlman and McColl, 1986). The greater ability of organic acids, compared with mineral acids of comparable strength, to release P from PR and the direct evidence of their chelating ability have been documented (Johnston, 1954a, 1954b; Kpomblekou and Tabatabai, 1994a). Another important factor in the release of P from PR is the participation of the OH groups in the organic acids. For example, it has been shown that citric acid with three carboxyl (COOH) groups and an OH group was able to dissolve more P from PR than cis-aconitic acid with three carboxyl groups but without the OH group (Kpomblekou and Tabatabai, 1994a).

Fulvic acid is the most reactive of the humic substances in adsorbing significant amounts of Ca^{2+} and releasing H^+ ions, thereby enhancing PR dissolution. Humic acid may form complexes with P and Ca, and create a sink for further dissolution of PR (Singh and Amberger, 1990). The application of humic substances to soil also makes more P available to plants by competing for, and by forming a protective coating over, soil phosphate-sorption sites. An additional benefit that accrues from the application of phospho-compost is the movement of dissolved P to a greater soil depth, which provides a larger soil volume for P uptake by plants.

Practical considerations of phospho-composting

Organic manure is a broad term that comprises: manures prepared from cattle dung, excreta of other animals, crop residues, rural and urban composts, and other animal wastes. The concentration of nutrients in organic materials is variable. Although most of these materials contain significant amounts of nitrogen (N), they contain little P (Table 28). The effectiveness of the composts in solubilizing PR varies with the kind and composition of waste material and with the rate of decomposition. It is a function of the magnitude of production of organic acids and chelating substances in the compost, which in turn result from the metabolic activities of micro-organisms including bacteria, fungi and actinomycetes. Reports indicate that plant materials

TABLE 28
Nutrient content of organic waste materials and farmyard compost (dry basis except for urine)

Category	Source	Nutrient content (%)		
		N	P	K
Animal wastes	Cow manure	0.7	0.5	2.31
	Cattle urine	0.8	<0.01	0.03
	Sheep and goat dung	2.0	0.51	2.32
	Night soil	1.2	0.35	0.21
	Leather waste	7.0	0.04	0.10
Farmyard manure	Farmyard manure	0.8	0.08	0.25
Composts & plant residues	Poultry manure	2.9	1.26	0.97
	Town compost	1.8	0.44	0.62
	Rural compost	0.8	0.09	0.21
	Rice straw	0.6	0.08	2.10
	Lawn clippings and leaves	3.0	0.30	3.50

such as chopped leaves, crop residues (e.g. cereal straw) and lawn clippings composted with animal wastes are favoured because they produce more organic acids and humic substances. Composting PR with poultry manure may not be a preferred option because poultry manures contain large amounts of calcium carbonate and other basic compounds that hinder PR dissolution (Mahimairaja *et al.*, 1995).

Subba Rao (1982b) has provided some practical hints for the phospho-composting of agricultural wastes:

- Collect plant wastes, such as cereal straw, stalks and leaves, and make the composting mass heterogeneous by mixing two or three types of materials.
- Shred the materials to smaller pieces, preferably to 50–70 mm in order to increase their surface area.
- Adjust the initial carbon-to-nitrogen (C:N) ratio to 30:35. A lower C:N ratio encourages ammonia volatilization while a higher ratio slows the rate of decomposition.
- Mix the agricultural waste with animal dung and soil in the ratio of 70:20:10.
- Mix with PR at a rate of up to 30 percent weight for weight.
- Maintain moisture at about 60 percent weight for weight and temperature at 50–60 °C.

The rate of decomposition of the organic materials can be accelerated by microbial inoculation of the compost (e.g. *Aspergillus* spp., *Penicillium* spp., *Trichoderma viride*, *Cellulomonas* spp. and *Cytophaga* spp.), and by the addition of energy sources (molasses) to the waste before composting (Singh and Amberger, 1991).

A ratio of four parts of organic waste material to one part of PR (dry-weight basis) seems to be an effective combination. Singh and Amberger (1991) investigated dissolution of two sedimentary PRs (Mussoorie and Hyperphos) applied separately. They incubated PR mixed with wheat straw at a ratio of 1:4, inoculated with soil and compost extract, and also adjusted the C:N ratio by adding urine. The authors reported that the dissolution of PR increased in 30 d to 7 percent of total P for Hyperphos PR and 15 percent for Mussoorie PR. The addition of molasses enhanced P solubility by a further 3 percent for both PRs.

Agronomic effectiveness of phospho-composts

An increase in the agronomic effectiveness of PR in phospho-composts over that of directly applied PR can be expected because of its greater water-soluble and citric-soluble P contents, which would be available to plants. Moreover, the soluble P fractions should stimulate root growth and facilitate greater exploitation of P enriched soil (Chien *et al.*, 1987a; Rajan and Watkinson, 1992; Habib *et al.*, 1999). Published work on the agronomic effectiveness of phospho-composts appears to be scarce.

Phospho-compost prepared by mixing farm wastes, cattle dung and soil (as a source of bacterial inoculum) has been found to be as good as single superphosphate (SSP) (Bangar *et al.*, 1985; Palaniappan and Natarajan, 1993). The P sources were applied on an equivalent total P basis to a tropical soil and the crops were: pigeon pea, green gram, clusterbean, wheat and pearl millet. At pH values higher than 7.5, where directly applied PR is not expected to dissolve, phospho-compost was as effective as SSP (Table 29) (Mishra and Bangar, 1986). However, the P sources were applied at a single rate of P in these studies, which reduces the value of the results.

TABLE 29
Effect of P sources on yield and P uptake for red gram and clusterbean

Treatment	Red gram yield		Clusterbean yield	
	tonnes/ha			
	Grain	Straw	Grain	Straw
Control	1.47	5.10	0.54	1.9
Mussoorie PR	1.57	5.40	0.68	2.16
Single superphosphate	1.64	5.65	0.82	2.53
Phospho-compost	1.83	6.15	0.84	2.63
Compost only	1.69	5.50		
Compost + Mussoorie PR	1.67	5.75		
LSD (p = 0.5)	0.14	0.67	0.18	0.38

Note: All P sources applied at equivalent P application rate of 17.3 kg/ha.

Future research needs

Phospho-composting offers the advantage of using otherwise unusable PRs, and the environmental advantage of safe disposal of organic waste. In situations where organic manures are already in use or are a viable alternative to chemical fertilizers (Mugwira *et al*., 2002), phospho-composting has an advantage. On the other hand, if phospho-compost is to be applied mainly as a source of P, then the benefits need to be weighed against the cost of preparation and application. Further research is needed to determine scientifically the minimum amount of compost required to solubilize PR to a level where the product would be economically as effective as water-soluble phosphate (WSP) fertilizers. The research programme should include PRs of different reactivities and various combinations of locally available organic wastes.

Inoculation of seedlings with endomycorrhizae

Fundamentals

The term mycorrhiza refers to the symbiotic association between plant roots and fungi. In nature, most plant roots form mycorrhizal associations of one kind or another with fungi in soil, with the mycorrhizal fungi performing the function of root hairs. The type of mycorrhiza that improves P uptake by plants is vesicular-arbuscular mycorrhizae (VAM), and the commonly used spore types are *Glomus fasciculatum*, *G. mosseae*, *G. etunicatum*, *G. tenue* and *Giaspora margarita*. VAM fungi infect the cells of the root cortex and form both an internal network of hyphae and an external growth of hyphae. They possess special structures known as vesicles and arbuscules. The highly branched arbuscules help in the transfer of nutrients from the fungus to the plant-root cells, and the vesicles are sac-like structures, which store P as phospholipids.

VAM are geographically ubiquitous and occur over a wide ecological range from aquatic to desert environments (Mosse *et al*., 1981; Bagyaraj, 1990). VAM fungi colonize families of most agricultural crops. Families that rarely form VAM include Cruciferae, Chenopodiaceae, Polygonaceae and Cyperaceae.

Miyasaka and Habte (2001) have reviewed the integration of arbuscular mycorrhizal fungi into cropping systems in order to maintain yields while reducing P inputs.

Mode of action

Enhanced P uptake in VAM-infected plants seems to be facilitated by: (i) the fungal hyphae exploring a greater volume of soil for P and also intercepting a greater number of point sources of P; (ii) the fungi dissolving sparingly soluble P minerals (e.g. PR); and (iii) the infected roots increasing the rate of P uptake, by increasing the diffusion gradient by depleting P to lower P concentrations than can non-mycorrhizal roots and by enhancing the transfer of P between living roots and from dying roots to living roots (Bolan and Robson, 1987; Sylvia, 1992; Frossard *et al.*, 1995; Lange Ness and Vlek, 2000; Brundrett, 2002). The P inflow rates of mycorrhizal roots are calculated to be 2–6 times those of non-mycorrhizal roots (Jones *et al.*, 1998).

The effect of mycorrhizal relative to non-mycorrhizal plants on the uptake of P varies with the P application rate, with uptake generally enhanced at low to intermediate levels of P application (Ortas *et al.*, 1996; Sari *et al.*, 2002). The effect of infection is to alter the P response curve such that it rises more steeply and reaches a maximum at much lower levels of P applied, either as soluble or sparingly soluble P as in PR. Although some researchers (Murdoch *et al.*, 1967) have concluded that the effectiveness is greater with low-solubility fertilizers (e.g. PR), such conclusions appear to be the result of procedures used with one or two rates of P application rather than several rates (Pairunan *et al.*, 1980). For the mycorrhiza to be effective, some threshold concentration of P in soil solution needs to be reached (Bolan and Robson, 1987) and this could be as low as 0.02 mg/litre (Manjunath and Habte 1992). At the same time, high levels of P in solution may decrease the level of mycorrhizal infection (Kucey *et al.*, 1989).

Application of technology to enhance PR availability

Compared with non-mycorrhizal plants, Pairunan *et al.* (1980) have reported that mycorrhizal plants enhance the effectiveness of fertilizers by about 30 percent, including that of PR. As mycorrhizal fungi already infect plants of most families, the aim should be to introduce the most efficient mycorrhizal endophyte into plants. Studies indicate that the introduction of appropriate VAM fungi can increase rice grain yields by three times compared with uninoculated treatments (Secilia and Bagyaraj, 1992). Arihara and Karasawa (2000) reported that, under field conditions, yield and P uptake by maize were enhanced when grown after the cultivation of mycorrhizal crops.

Various methods for introducing VAM inoculums into a field crop have been examined, e.g. coating seeds with VAM inoculums and placing inoculums in the field under seeds in furrows. It has been found that the most attractive and practical way to use VAM inoculums is with vegetatively propagated crops and/or transplanted crops. In this situation, growers need to incorporate inoculums in the seedling trays or nursery beds. Seedlings thus raised will be colonized by the introduced fungi, which can then be transplanted out in the field. This method has been used successfully in India on agronomically important crops (e.g. chilli, finger millet, tomato, citrus and mango) and on forest tree species (e.g. *Tamarindus indica* and *Acacia nilotica*). VAM fungi are commercially available in some countries such as India. VAM inoculation for raising citrus seedlings is in commercial use in the United States of America (Menge *et al.*, 1977).

Future research

There is strong evidence that VAM fungi increase the uptake of P by plants from PRs (Pairunan *et al.*, 1980; Barea *et al.*, 1983; Toro *et al.*, 1997; Yusdar and Hanafi, 2003). This technology is particularly promising with crops raised in nurseries, seedling trays or polythene bags and then planted out in the field. Further research is needed to integrate VAM fungi into the cropping system with the specific aim of using PR as a phosphate fertilizer. It needs to assess the quantitative effects of chosen strains of fungi on P uptake from PRs by crops relative to that in non-inoculated natural soils. Such studies should include: PRs of differing reactivities; a WSP fertilizer (e.g. SSP) as the reference product; and several rates of P application so that response curves can be drawn. The studies should be conducted under field conditions so that the cost-benefit ratio of the use of PR with and without VAM fungi in relation to the use of WSP fertilizers can be calculated. As there is evidence that VAM fungal species are highly host specific and variable in their response to the mineral environment of soils (Bever *et al.*, 2001; Ortas *et al.*, 2002), the research programmes should consider these parameters.

Use of ectomycorrhizae

Ectomycorrhizae are generally found on woody perennials in families such as Betulaceae, Dipterocarpaceae, Fagaceae, Myrtaceae, Pinaceae and Salicaceae. The ectomycorrhizae are characterized as a root-fungus association, in which the fungus grows as a sheath or mantle on the surface of the roots. The net of hyphae extends into the cortex of plants but is confined to the intercellular spaces (unlike VAM), which form intracellular structures and produce an interconnection network known as a Hartig net. From the mantle, the hyphae extend into the soil and enhance the transport of phosphate and water to the host plants (Duddridge *et al.*, 1980). This should assist in enhancing the availability of P from slow-release fertilizers such as PR. The fungal partner of ectomycorrhizae can be grown on synthetic medium for inoculation. Castellano and Molina (1989) discuss detailed procedures for inoculum production and nursery inoculation.

Use of phosphate solubilizing micro-organisms

A group of heterotrophic micro-organisms have been reported to solubilize inorganic forms of P. This is achieved by excreting organic acids that dissolve phosphatic minerals and/or chelate cationic partners of the P ion directly, releasing P into solution (Halder *et al.*, 1990; Gaur, 1990; Bojinova *et al.*, 1997; He *et al.*, 2002). Some important micro-organisms include the bacteria *Bacillus megaterium*, *B. circulans*, *B. subtilis*, *B. polymyxa* and *Pseudomonas straita*. Fungal micro-organisms include *Aspergillus awamori*, *Penicillum bilaii*, *P. digitatum* and *Trichoderma* sp. Analyses of culture filtrates have identified a number of organic acids such as lactic, glycolic, citric, 2-ketogluconic, malic, oxalic, malonic, tartaric and succinic acids, all of which have chelating properties (Kucey *et al.*, 1989).

Field trials in India and the then USSR have shown that the use of phosphate solubilizing micro-organisms (PSMs) can increase crop yields by up to 70 percent (Verma, 1993; Wani and Lee, 1992; Subba Rao, 1982a). The crops include oats, mustard, sugar beets, cabbage, tomato, barley, Egyptian clover, maize, potato, red gram, rice, chickpea, soybean and groundnut. *In vitro* studies have demonstrated the dissolution of PR by PSMs (Barea *et al.*, 1983). Results from greenhouse trials have indicated a greater response of wheat and onion to PR application when seeds or seedlings are inoculated with PSMs. The increase in growth is greater with VAM fungi and PSMs in combination than when these organisms are used singly (Young, 1990; Toro *et al.*, 1997; Singh and Kapoor, 1999). It is likely that PSMs dissolve sparingly soluble P, which is taken up by VAM mycelia, by more than one process, including the release of organic acids (Illmer *et al.*, 1995) and the solubilization of calcium phosphates (Illmer and Scinner, 1995).

Research needs

There is strong evidence that PSMs, especially in combination with VAM, increase the agronomic effectiveness of PRs (Barea *et al.*, 2002). PSMs are commercially available in some countries (e.g. India), where large-scale cultivation of PSMs is underway (by growing cultures in large flasks on rotary shakers or in batch fermentors). However, it is not evident whether these micro-organisms can increase the effectiveness of PRs under standard field conditions to such a magnitude that PR can be used as an alternative fertilizer. Field experiments need to be set up at specific sites in order to provide rigorous scientific testing of the practical usefulness of PSMs and VAM in enhancing the effectiveness of PRs, and thereby promoting their increased use.

Use of plant genotypes

P-uptake efficiency can be enhanced by selecting plant species or genotypes that exhibit several mechanisms for increased P absorption under low P conditions, such as: (i) greater root growth and, thus, increased extent of soil exploration; (ii) the exuding of proton and organic acids that increase the solubility of sparingly soluble phosphates by decreasing pH and/or chelation; and (iii) the production of elevated levels of phosphatase enzymes that can break down organic P to inorganic P (Miyasaka and Habte, 2001). Another approach is to use plants that are tolerant to Al toxicity (Ishikawa *et al.*, 2000).

Increased soil acidity in the rhizosphere can enhance PR dissolution and its availability to plants. This has been observed directly as increased PR dissolution but more often indirectly as increased P uptake by those plants that acidify the rhizosphere (Bekele *et al.*, 1983; Hedley *et al.*, 1983; Moorby *et al.*, 1988; Gahoonia *et al.*, 1992; Haynes, 1992; Nakamaru *et al.*, 2000). Hinsinger and Gilkes (1995) and Habib *et al.* (1999) have also found enhanced PR dissolution by the rhizosphere of some crop species in alkaline soils. Proton secretion by roots occurs when the equivalent sum of cation uptake by plants (K^+, Ca^{2+}, Mg^{2+} and Na^+) exceeds that for anions (usually, NO_3^-, $H_2PO_4^-$, SO_4^{2-}, and Cl^-). Soil acidification in the rhizosphere is greater for N-fixing legumes, which accumulate N as NH_3 through symbiosis with *Rhizobium* micro-organisms. For this reason, leguminous plants are particularly suited for PR use.

The roots of some plants (e.g. rape) may also enhance PR dissolution by secreting organic acids, such as malic, citric, oxalic and 2-ketogluconic acids, that can be expected to complex cations of PR (Ca, Al and Fe) in addition to lowering rhizosphere soil pH (Moghimi and Tate, 1978; Hoffland *et al.*, 1989; Zapata *et al.*, 1996; Nakamaru *et al.*, 2000; Montenegro and Zapata, 2002).

Strategies to improve PR effectiveness involve the use of crop genotypes with increased acidity in the rhizosphere, such as N-fixing and acid-secreting legumes as intercrops (grain legumes) or mixed swards (pastures), which will enhance PR dissolution and make P available to nearby crop plants. An alternative is to include these crops in the crop rotation so that additional PR dissolves and enters the soil labile pool P.

A new strategy is to improve existing plant species genetically in order to increase root secretion of organic acids and protons. De la Fuente *et al.* (1997) reported that increased secretion of citrate occurred from roots of tobacco as a result of insertion of a citrate synthase gene from *Pseudomonas aeruginosa*. More research needs to focus on the possibility of inserting organic acid-producing genes into plants, which are poor utilizers of PR. Recently, a combined approach, i.e. to search for Al-tolerant and P-efficient genotypes, has been proposed in order to develop sustainable cropping systems for the acid soils of the tropics and subtropics (Hocking, 2001; Keerthisinghe *et al.*, 2001).

CHEMICAL MEANS

Partial acidulation of phosphate rock

Partially acidulated phosphate rocks (PAPRs) are prepared by reacting PRs, usually with H_2SO_4 or H_3PO_4, in amounts less than that needed to make SSP or triple supersphosphate (TSP), respectively. The use of PAPRs has been widespread in Europe and South America since Nordengren (1957) first reported their use. PAPRs may offer an economic means of enhancing the agronomic effectiveness of indigenous PR sources that may otherwise be unsuited for direct application. For this reason, extensive studies have been, and continue to be, conducted internationally (Hammond et al., 1986b; Rajan and Marwaha, 1993; Chien and Menon, 1995a; Chien, 2003; Zapata, 2003). PAPRs are cheaper than fully acidulated WSP fertilizers because less acid and energy is required per unit of P in the product. In addition, PAPRs are often more concentrated than SSP. Thus, in some situations partial acidulation may be a preferred way of improving the effectiveness of imported PRs.

Choice of level and acid for partial acidulation

The level of acidulation of PR is usually referred to in percentage terms. For example, when one-fifth of the sulphuric acid needed to make superphosphate from that particular PR is used, the product is referred to as PAPR-20 percent H_2SO_4. The term PAPR-20 percent H_3PO_4 refers to phosphoric-acid-based PAPR using 20 percent of the H_3PO_4 required for TSP.

Figure 31 illustrates the relationships between the level of acidulation of a reactive PR (North Carolina PR) with H_3PO_4 or H_2SO_4 and the total and water-soluble P of the product. As H_3PO_4 contains water-soluble P, partial acidulation of PR with H_3PO_4 always results in PAPRs containing more total and water-soluble P than the unacidulated PR. Partial acidulation with H_2SO_4 results in a decrease in total P because of the formation of calcium sulphate in the product. However, the water-soluble P increases with the increasing degree of acidulation. The slope of the water-soluble P curve is smaller initially because of the presence of accessory minerals such as free carbonates, which are more soluble than apatite and preferentially consume the acid added. It is possible to prepare PAPRs from PRs of low reactivity. However, if the low reactivity is due to high contents of iron and aluminium oxides, the PR may not be suitable for partial acidulation with H_2SO_4. This is because the water-soluble P formed reverts to water-insoluble P during the preparation of PAPR (Hammond et al., 1989).

Ideally, partial acidulation of a given PR should be decided after a development study that varies: (i) the type of acid such as sulphuric, phosphoric and mixed acids and a mixture of sulphuric acid and ammoniacal salts; (ii) the quantity and concentration of acids; (iii) the temperature; and (iv) the type of mixing procedure (e.g. adding PR to bulk acid with constant mixing or spraying acid as fine droplets to a falling curtain of PR particles). The objective is to solubilize the maximum amount of PR for a given level of acid addition. Experience has shown that the most effective levels of acidulation, which give maximum soluble P for the quantity of acid used, range from 30 to 60 percent. At these rates, the acids affect the apatite but have little effect on hard gangues such as silicates.

The major P components in PAPR products are water-soluble monocalcium phosphate and sparingly soluble acid-unreacted PR. Partial acidulation and granulation of PR with H_3PO_4 yields products containing a soluble binder, monocalcium phosphate, that readily dissolves on application of the fertilizer to soil, so allowing the dissolution of the residual PR. On the other hand, partial acidulation and granulation with H_2SO_4 may yield products with $CaSO_4$ coating

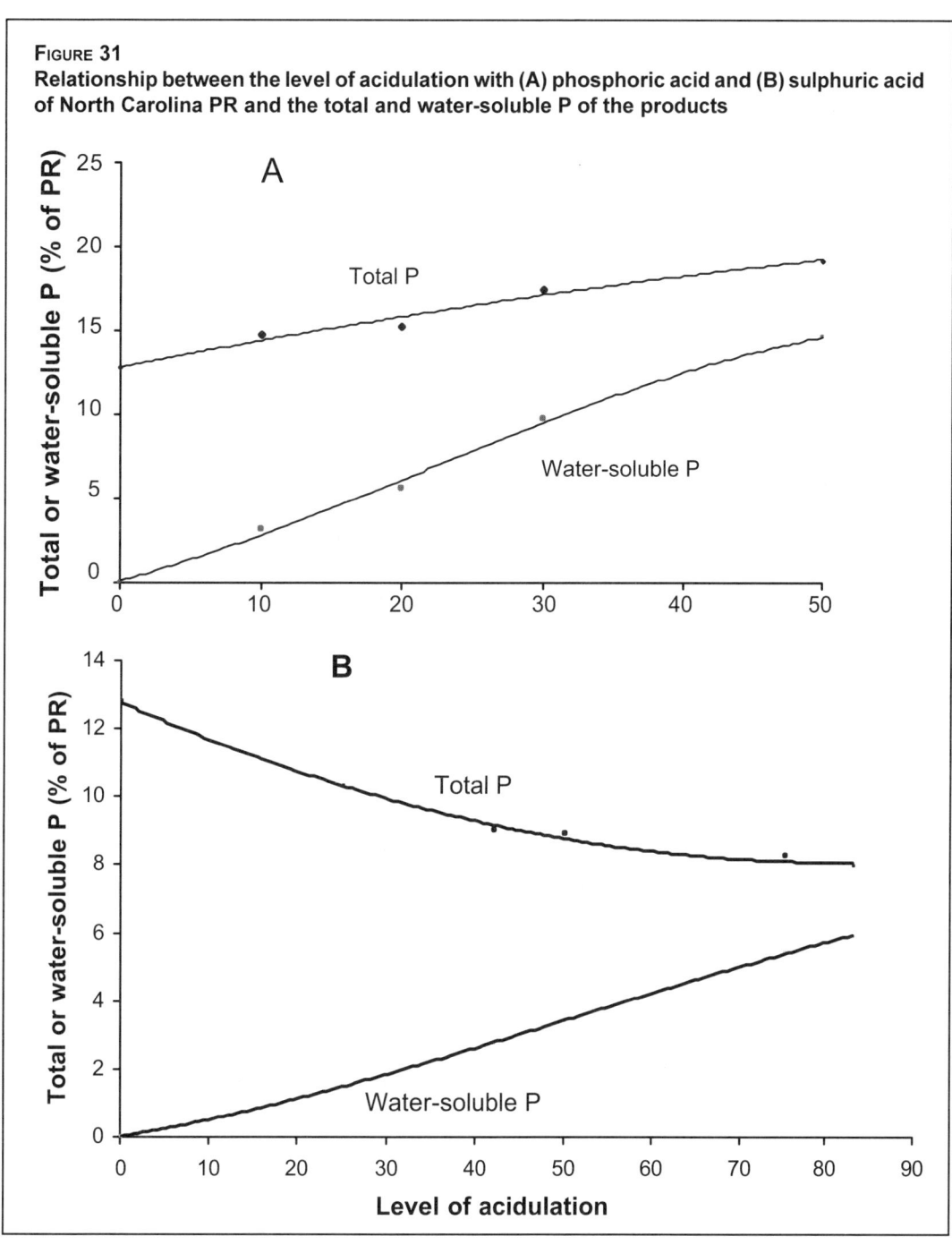

FIGURE 31
Relationship between the level of acidulation with (A) phosphoric acid and (B) sulphuric acid of North Carolina PR and the total and water-soluble P of the products

PR particles. This may hinder the dissolution of the residual PR (Rajan and Ghani, 1997). It is possible to overcome this problem by controlling the drying temperature, as demonstrated by researchers of the International Fertilizer Development Center (IFDC) by preparing PAPR in a single-step process (Schultz, 1986).

The agronomic performance of PAPRs

The agronomic performance of PAPRs varies with: (i) the physical and chemical properties of the PR used; (ii) the degree of acidulation; (iii) the soil chemical properties, especially soil pH and P sorption (or retention); and (iv) the cropping system. In general, factors that enhance the effectiveness of PAPRs are: high reactivity and fine particle size of the PR, acidic soils, high soil P retention, long growth cycle, and crop rotation. The product is also likely to be more beneficial than water-soluble P fertilizers in situations where phosphate leaching may be a problem, as in sandy soils (Bolland *et al.*, 1995). Increasing the level of acidulation increases the water-soluble P content and the agronomic effectiveness of PAPRs, but this needs to be weighed against the increased cost of fertilizer production. In all cases, the agronomic performance of the products produced should be evaluated to determine the cost-effective level of acidulation for a set of conditions (Owusu-Bennoah *et al.*, 2002; Casanova *et al.*, 2002a; Rodriguez and Herrera, 2002).

Research on the agronomic effectiveness of PAPRs has been reviewed (Bolan *et al.*, 1990; Hammond *et al.*, 1986b; Rajan and Marwaha, 1993; Chien and Menon, 1995a; Chien, 2003) and the general conclusion is that PAPRs of 40–50 percent acidulation with H_2SO_4 or 20–30 percent acidulation with H_3PO_4 are as effective as fully acidulated superphosphate. In soils of a high pH (pH 6.5–8.0), PAPRs may be as effective as superphosphate, albeit at a higher degree of acidulation with H_3PO_4 of up to 50-percent acidulation (about 66 percent of total P in water-soluble form) (McLay *et al.*, 2000; Chien 2003).

The possible explanations for the high agronomic effectiveness of PAPRs are: (i) root priming effect by the monocalcium phosphate component of PAPR and the subsequent greater exploitation by roots of P-enriched soil; and (ii) dissolution of the monocalcium phosphate leading to formation of H_3PO_4 and the reaction of the acid on the unreacted PR (secondary acidulation). Root-growth studies have demonstrated a marked increase in root development in PAPR-implanted compared with PR-implanted soil samples (Rajan and Watkinson, 1992). A greater dissolution of the residual PR in PAPRs than unacidulated PR has also been documented under both greenhouse and field conditions (Rajan and Watkinson, 1992; Rajan and Ghani, 1997; McLay *et al.*, 2000).

PAPR manufacture in an SSP plant

PAPRs have been manufactured in New Zealand at an SSP plant by mixing reactive PR with SSP (30:70 ratio by weight) at the ex-den stage, when the SSP was still wet. The product traded sold under the name Longlife Super. The extent to which the PR was partially acidulated depended on the acid-PR ratio used to make the SSP and, consequently, on the amount of free acid that was available to react with the PR (Bolan *et al.*, 1987; Hedley *et al.*, 1988). The water-soluble P was contributed mostly by the SSP component and to a minor extent by the partial acidulation of the PR added.

The agronomic effectiveness of Longlife Super applied to permanent pastures was determined in six field trials over three years (Ledgard *et al.*, 1992). The fertilizers contained 35-50 percent of their total P in a water-soluble form and the soil pH values were about 5.7. The results showed that the pasture response to Longlife Super was generally less than that for superphosphate. Systematic studies indicated that the poor performance of Longlife Super was due to the $CaSO_4$ continuing to cement PR particles as granules for six weeks, thereby depressing their dissolution.

Physical means

Compacted PR with water-soluble phosphate products

Fertilizers that are similar in chemical composition to PAPRs can be prepared indirectly by compacting dry PR with soluble phosphate fertilizers, such as SSP or TSP, under pressure (Chien *et al.*, 1987a; Chien and Menon, 1995a; Menon and Chien, 1996). The water-soluble P content of the products will depend on the ratio of PR to WSP fertilizer used. Compaction technology offers the advantage of using PRs that are not suitable for direct partial acidulation with H_2SO_4 because of their high Al and Fe sesquioxides content. There is evidence that the effectiveness of PRs, even those of low reactivity, is increased after compacting with water-soluble P (Chien *et al.*, 1987a; Kpomblekou *et al.*, 1991). Under these conditions, compaction of PR with water-soluble P fertilizers at P ratios of about 50:50 can make the use of indigenous PRs agronomically and economically attractive in developing countries. However, the agronomic effectiveness of compacted fertilizer products relative to water-soluble fertilizers will depend on a number of factors, in a similar way to PAPRs.

Dry mixtures of PR with water-soluble phosphate fertilizers

There is field trial evidence that the application of PR as a dry mixture with WSP fertilizers can increase the effectiveness of the PR applied. Research conducted on an Indian calcareous soil indicated that applying fertilizers at a single rate in a 1:1 mixture of Mussoorie PR and SSP could be as effective as SSP (Siddique *et al.*, 1986). In another field study, Mussoorie PR and SSP were applied at a P ratio of 2.2:1 using three fertilizer application rates (Singaram *et al.*, 1995). The results showed that in an alkaline soil (pH 8.2) the PR-SSP mixture tended to be as effective as SSP. Moreover, based on a sequence of three crops, the product was economically equal to SSP. It was calculated that the dissolution of PR increased by 55 percent when applied in combination with SSP compared with that of an application of PR only.

In a greenhouse study, Chien *et al.* (1996) made quantitative estimates of the enhancing influence of water-soluble P (in the form of TSP) on the agronomic effectiveness of a medium-reactive central Florida PR using radioactive P tagged (^{32}P) fertilizers. At a P ratio of 1:1, the PR and TSP were applied separately to the soil. Based on P uptake by plants, the study reported that the effectiveness of PR in combination with TSP increased by 165 percent for maize and 72 percent for cowpea. In another greenhouse study, Habib *et al.* (1999) found that a physical mixture of PR and TSP was as effective as TSP for growing rape in a limed alkaline soil. Zapata and Zaharah (2002) also found an enhancement effect on the agronomic effectiveness of Florida PR and sewage sludge when applied in mixture with TSP.

The increased effectiveness of PR when applied as a dry mixture can be ascribed to the root-priming effect of soluble P and, therefore, increased root exploitation of the PR added. Considering the negligible processing input and the resulting improvement in PR effectiveness, the above findings are of significant economic importance. Further research is needed under field conditions in order to evaluate the usefulness of this process.

Phosphate rock elemental sulphur assemblages

Numerous studies have shown that the agronomic effectiveness of PRs can be improved when they are applied after admixing or cogranulating with sulphur (S). In some instances, the products were inoculated with sulphur-oxidizing bacteria *Thiobacillus* spp. Thus, the product was referred to as 'biosuper' (Swaby, 1975). Kucey *et al.* (1989) reviewed the role of microbes in increasing plant-available P.

The principle behind the use of phosphate rock elemental sulphur assemblages (PR/S) is that the inoculated or native population of soil bacteria oxidizes S to H_2SO_4 when the product is applied to the soil. This acid in turn reacts with the PR particles that are in close proximity to S and forms monocalcium and dicalcium phosphates. Thus, the dissolution of PRs in soil is assisted by localized acidulation, in addition to that caused by ambient soil acidity. The important sulphur-oxidizing bacterial species are *Thiobacillus thioxidans* and *Thiobacillus thioparus*. Inoculating soils already rich in *Thiobacillus* spp. may not be essential (Rajan, 1982a) but inoculation is preferred for speedy multiplication of the bacteria and PR dissolution after application to soil.

PR/Ss are attractive because: (i) the production is not capital intensive; (ii) they enable flexible PR:S ratios; (iii) low-grade PRs that may be unsuitable for making soluble fertilizers can be used; and (iv) they behave as controlled-release P and S fertilizers. On the negative side, for speedy oxidation of S, the S needs to be present as fine particles (95 percent minus 0.15 mm), either ground with PR or ground separately and mixed. Grinding S without admixing PR or in a dry state, without moisture addition, is a fire hazard and it is not encouraged. Nevertheless, in New Zealand, a pilot plant has recently produced PR/Ss containing *Thiobacillus* spp. bacteria (EnviroPhos™) using a cost-effective technique.

Factors that influence the effectiveness of PR/Ss are: (i) reactivity of the PR; (ii) PR:S ratio; (iii) type of crops; and (iv) soil environment. The agronomic effectiveness of PRs applied as PR/S can be expected to increase with either increasing reactivity of PRs or with increased fineness of PR particles (Rajan, 1982b; Loganathan *et al.*, 1994). Research shows that the PR/S prepared from a finely ground reactive North Carolina PR, at a PR:S ratio of 5:1, was as effective as SSP, whereas that prepared from an unreactive Florida PR was inferior. Nevertheless, there was an increase in the agronomic effectiveness of both PRs when they were applied as PR/S, the values being 18-30 percent for the North Carolina PR and 50-70 percent for the Florida PR depending on the rate of application (Rajan, 1982b).

As the H_2SO_4 produced on oxidation of S affects the dissolution of PR in PR/Ss, it is logical to conclude that increasing the S content in PR/Ss would increase their agronomic effectiveness. On the other hand, a balanced approach is needed as the greater the enrichment of the product with S, the greater will be the cost of the product. Results from early studies show that PR/Ss could be as effective as SSP when the PR:S ratios are between 1:1 and 5:1 (Kittams and Attoe, 1965; Attoe and Olson, 1966; Swaby, 1975). The 5:1 ratio is similar to the proportion in which PR and S in the form of H_2SO_4 are used for making SSP. However, mostly unreactive PRs such as Florida PR (the United States of America) and Queensland PR (Australia) were used. When finely ground reactive PRs are used, this ratio can be increased to 7:1 without losing agronomic effectiveness (Rajan, 1983). On long-term crops such as permanent pastures, PR/Ss prepared at a PR:S ratio of 14:1 can be agronomically as effective as superphosphate (Rajan, 2002). On the other hand, for short-term crops that require a high rate of P supply, the PR/Ss prepared from reactive PRs with a narrow PR:S ratio may be required. As for the influence of soil environment, the agronomic effectiveness of PR/Ss will be greater in sites that favour PR dissolution. Under such conditions, PR/Ss prepared from less reactive rocks, or with wide PR:S ratios if the PRs are highly reactive, can be used as effectively as phosphate fertilizers.

The effectiveness of S will be highest if the S particles are in intimate contact with PRs as this will facilitate the maximum reaction of the H_2SO_4 produced on the PR. For this reason,

PR/Ss have been obtained by cogranulating PR with elemental S (Swaby, 1975; Rajan, 1982b). On the negative side, granulation restricts the surface area for soil acid to react on the PR. For this reason, granules of less than 2 mm are preferred. Moist physical mixtures of PR/S containing *Thiobacillus* spp. have been found to be as effective as SSP for pasture production (Rajan, 2002).

Chapter 10
Economic factors in the adoption and utilization of phosphate rocks

Increased population pressure, reduced length of fallow, deforestation and improper agricultural practices have led to widespread soil degradation in many parts of the developing world. An important manifestation of this environmental damage is the inadequate replenishment of soil nutrients and organic matter. In particular, phosphorus (P) deficiency is becoming critical in many soils. Moreover, because of complementarities in the uptake of plant nutrients, this deficiency threatens to disturb the viability of applying other nutrients. In order to preserve the sustainability of agriculture and safeguard the livelihood of large segments of the rural population, there is an urgent need to rebuild soil fertility and so maintain and improve current levels of productivity and farm income (Heerink *et al.*, 2001).

Sustainable intensification and shifts towards higher-value crops require a careful application of external inputs such as inorganic fertilizers. A number of factors have constrained fertilizer use, especially in sub-Saharan Africa and low-income Asia. The most important of these factors are: the limited financial means and risk-taking capacity of farmers; poor and expensive distribution systems for fertilizers (and marketable crop surpluses); a lack of adequate knowledge of the potential of using local phosphate rock (PR); and the absence of non-industrial techniques to increase the solubility of PR (Appleton, 2001).

In many countries, foreign-exchange shortages have constrained fertilizer use by restricting imports. Moreover, domestic supply has suffered from government interventions and regulations that have hampered the emergence of a private fertilizer trade. As a result, these countries have become excessively dependent on fertilizer aid. As most bilateral donors are reluctant to make long-term commitments for fertilizer aid, dependence on aid introduces a high degree of uncertainty and prevents the development of efficient input-supply and marketing systems. It also discourages fertilizer use because farmers do not feel secure in adopting fertilizer-intensive cropping practices when access to fertilizer is uncertain. Another disadvantage of fertilizer aid is that cheap, subsidized fertilizer aid reduces incentives to develop domestic resources, especially PR for direct application.

Following the implementation of structural adjustment programmes, the real price of imported fertilizer increased substantially, creating a further constraint on its use. However, at the same time, more expensive imported fertilizer is an incentive to develop domestic substitutes. Coupled with a generalized donor fatigue with project and commodity aid, the restoration of soil fertility through programme aid is being considered increasingly as a lasting and more cost-effective substitute for food aid. Through higher yields, improved soil fertility constitutes a long-term contribution to enhanced food security.

Expenditure on the application of P sources, especially those with a significant content of less soluble P such as PR, can be considered as a restoration of the natural resource base. This is because it augments and maintains the stock of natural capital embodied in soil resources.

In this respect, PR is to be seen as an amendment that improves the soil, making the use of current inputs such as fertilizer for crop uptake, water and labour more efficient. Both farmers and society benefit from increasing agricultural output and decreasing nutrient depletion in agricultural production (Gerner and Baanante 1995). Defined in this way, PR application can be considered a capital investment.

The distinction between PR application as a soil amendment and as a crop fertilizer is an important one. Phosphate fertilizer can be produced through relatively simple PR beneficiation as well as through further industrial processing. PR and water-soluble phosphates, such as single superphosphate (SSP), triple superphosphate (TSP) and di-ammonium phosphate (DAP), are then competing inputs that farmers will select on the basis of price, availability, quality and other characteristics. The relevant question is under which circumstances PR based on domestic resources has a competitive edge over industrial fertilizers. Moreover, farmers can combine PR as a basal soil amendment with regular applications of industrial fertilizers.

PR as a capital investment raises various issues regarding the financing of such soil amendments. First, investment is lumpy, requiring an initial investment of large doses of PR that yields returns over a long period of time. Resource-poor farmers often do not have adequate financial resources to invest, nor are commercial banks willing to lend because of the high risk and poor collateral. Second, it rebuilds the natural fertility of depleted soils. Investment in soils is comparable to land reclamation or land improvement projects. They all require public investment, which enhances private profitability on a long-term basis and changes comparative advantages in production.

Possible adoption and use of phosphate rock by end users

Case studies of PR application in Burkina Faso, Madagascar, Mali and Zimbabwe (World Bank, 1997; NEI, 1998; Kuyvenhoven *et al.*, 1998b; Henao and Baanante, 1999) show that farm-level rates of return can be highly attractive for various crops, and even more so when the environmental impacts on society at large are included. Despite these results, adoption in practice remains limited. In contrast with local PR, derived and usually imported phosphate fertilizers often perform better (Appleton, 2001). However, widespread use of such imported sources of P cannot be observed either, except for a number of commercial crops. A cost-benefit analysis of fertilizer application captures only part of farmers' adoption behaviour. In order to understand the factors that constrain adoption more fully, a more complete analysis of farm-household livelihood strategies is needed.

In general, farm households adopt those practices, technologies, enterprises and activities (agricultural and non-agricultural) that suit their objectives best given their own resource endowments, constraints and the socio-economic environment in which they operate. This implies that particular interventions such as PR application should not be analysed in isolation but considered in conjunction with competing options (in terms of time, labour, finance, income, etc.) for the farm household or firm. It is in this sense that profitability is only a necessary condition for adoption.

Major factors that have been found to determine adoption, and that of PR in particular, are: ownership and property rights (or continuous possession) in order to ensure access to income flows; farm size; share of land under cropping and stocking rate (indicators of resource pressure); access to credit; off-farm income available for on-farm investment; farming system; the prices of and the access to timely and adequate supplies of fertilizers and other variable inputs; and knowledge of and access to information about the use of fertilizers and agricultural production technology in general. These factors influence the time horizon (and implicitly the discount rate) for investment decisions and the degree of relative risk aversion of the farm households involved.

Farmers often do not own the land they work. Therefore, they are often unwilling to invest in long-term land improvement. Farmers with well-established land access rights will obtain all of the short and long-term added revenues associated with PR application. Farmers who have usufruct of land on a long-term basis through land-leasing arrangements, such as sharecropping, will share these benefits with landowners who may be private landlords or with the community as a whole. Tenant farmers who are sharecroppers on the land for only one cropping season will not obtain any of the long-term benefits and only a part of the short-term benefits (Gerner and Baanante, 1995). The ownership/property rights land system in many of the least-developed countries (LDCs) can be characterized as a traditional communal type of land tenure and land use with the following characteristics: (i) low property concentration (sovereign rights vested in the community); (ii) decentralized cultivation (usufruct rights for group members); and (iii) production for subsistence. Hence, investment in PR is most likely to be implemented as a public action investment. Therefore, the community as a whole and not individual farmers should pay for it.

A capital investment in the soil by means of PR application has high up-front cost and yields returns over a long period of time. Access to credit is generally limited. However, off-farm income can help to overcome a capital constraint or may finance the purchase of a fixed-investment type of innovation (Feder *et al.*, 1985). Farm size can have different effects on the rate of adoption of new technologies, such as PR use, depending on the characteristics of the technology and institutional setting. The relationship between farm size and adoption depends on factors such as fixed adoption costs, risk preferences, human capital, credit constraints, labour requirements and tenure arrangements. An often-mentioned constraint on technology adoption by smaller farms relates to the fixed costs of implementation. In the case of PR application, these costs are of considerable importance. Large fixed costs reduce the tendency to adopt and slow the rate of adoption by smaller farms (Feder *et al.*, 1985).

Most farmers willing to invest in PR have a relatively intensive farming system. In such cropping systems, animal traction is already available and manure is applied. The PR to be applied to restore long-term fertility should also be available to farmers on time. This is because application after the optimal period reduces the possibility of immediate yield increases. Labour availability is another variable that affects farmers' decisions about adopting new agricultural practices or inputs. It is a particularly important factor in Africa (Helleiner, 1975). New technologies such as PR application increase the seasonal demand for labour. Hence, adoption is less attractive for farm households with limited family labour or those operating in areas with less access to labour markets. Labour shortages arise mainly as a result of male labour migration, the excessive burden on women and a lack of animal power. The consequences are late ploughing, late planting, late weeding, late harvesting and low yields.

A constraint that relates specifically to phosphatic fertilizer involves the time lag between application and visible effects. Depending on the reactivity of the PR, observable effects may occur during the cropping season of application or three or more years following application. Lack of education and extension services can be another constraint on PR use. The inability to read and understand fertilizer packages and instructions restricts the effectiveness of using written information as a means of disseminating knowledge about fertilizers. Field research experience has shown at least one unique constraint on the use of PR. In West Africa, farmers complained that when broadcasting finely ground PR, windy conditions forced the product into their eyes and caused a burning sensation.

The factors mentioned here correspond well with those given by Henao and Baanante (1999) in their Mali case study. They emphasize the "magnitude and relative importance of the marketable surpluses of rural households from farming, e.g. outputs and prices of crop and animal production. Sustainable expansion of farmers' marketable surpluses should be associated with expansion in the adoption of PR and other external inputs. The lack of marketable surplus associated with subsistence farming is a serious constraint to the adoption of PR and other external inputs. It is important to note that the sustainable growth of farmers' marketable surplus also depends on the existence of market outlets and the demand for their products that can ensure stable and suitable prices" (Henao and Baanante, 1999). Appleton (2001) reported similar conclusions on farmers' behaviour.

Enyong *et al.* (1999) examined farmers' perceptions and attitudes towards introduced soil-fertility enhancing technologies (SFETs) in West Africa. They concluded that farmers are knowledgeable about and practise SFETs that encompass PR application, crop residue and farmyard manure, chemical fertilizer and crop rotation to combat soil fertility decline. A number of factors influence their attitudes to and rationales behind adoption decisions. These factors include land use policies, labour resources, food-security concerns, perceived profitability, contribution to sustainability and access to information. Enyong *et al.* (1999) observed that some of these factors are beyond farmers' control and require a broad and integrated effort from research, extension and government to promote the use of the SFETs (including PR) in the region.

Cost of production, transport and distribution

Various countries in the developing world possess PR reserves, but few of these resources are exploited commercially on a substantial scale. Appleton (2001) provides a comprehensive survey on reserves, production and infrastructure conditions in sub-Saharan Africa, but warns that cost figures for PR production, transport and distribution vary widely, depending on site characteristics and available infrastructure. Each case is likely to be substantially different, precluding any general conclusion on the viability of mines and processing plants without a detailed technical, economic and institutional/organizational appraisal. Once such prefeasibility studies have justified further investigation, a full feasibility study can begin.

In view of the widely varying mining, production and distribution circumstances, this first section is confined to a brief overview. Subsequent sections provide some general information on the marketing of PR in various parts of the world, while the final sections present a case study on large-scale PR production in Venezuela to illustrate some characteristics of the production chain. The remainder of this section deals with aspects of PR production in sub-Saharan Africa.

Senegal and Togo produce PR, but the bulk of their phosphate is for the production of chemical fertilizer. Mali, Burkina Faso and Niger also have PR, but its exploitation is on a small scale. The quality is insufficient for use in the chemical industry. Moreover, transport costs are prohibitive. Several attempts have been made to promote large-scale utilization of PR, but success has been limited. One of the main reasons is the relatively high price of the product.

Equipment for grinding is rather heavy and investments are only profitable for large quantities. The equipment is technically simple and operational costs per tonne are relatively low. Enlarging the quantities would enable considerable economies of scale. While the existing factory in Burkina Faso produces some 2 000 tonnes of PR per year at an ex-factory price of about US$97.00/tonne, investment in a new factory to enlarge production to 30 000 tonnes/year would result in a price of less than US$24.25/tonne. This is comparable with the ex-factory prices of PR in Togo and Senegal (NEI, 1998).

PR is a voluminous product. Its bulky nature, combined with the long distances from the mines to the fields and the often poor conditions of the roads, makes unit transport costs very high. Where the production costs of PR are considerably lower than those of chemical fertilizers, this cost advantage is partly offset by the higher transport costs. This is because of the lower effective P content of PR in comparison with water-soluble P fertilizer and, hence, the larger quantities needed to obtain the same results. The same is the case for PR of different origin. Locally produced PR is cheaper than imported PR, but the advantage quickly diminishes with geographical distance. Consequently, PR application will remain limited geographically, but within such areas it might be interesting.

Unlike chemical phosphates, PR is not a homogenous product. First, the P_2O_5 content differs. Senegalese and Togolese PR have a content of some 38 percent, Malinese PR from Tilemsi has a P_2O_5 content of 32 percent and Burkinese PR a P_2O_5 content of about 27 percent P_2O_5. Another difference is the solubility. While TSP is water soluble, PR is water insoluble. In particular, the reactivity of Burkinese PR is relatively low. Consequently, more PR is needed in order to achieve the same fertilizer effect than a comparable quantity of TSP. On average, it is necessary to apply three to four times as much Senegalese PR compared with TSP and six times as much Burkinese PR in order to obtain the same fertilizing effect, at least in the short term. This is offset partly by the fact that the less soluble products have a more lasting effect, up to 7–10 years, but exact figures are not available. However, such a period exceeds the time horizon of most West African smallholders.

The differences in PR quality diminish the geographical area in which a certain PR product quality can be applied profitably. In Burkina Faso, unit prices of Burkinese PR are lower than those of imported PR. However, the importation of Senegalese PR might be more attractive in the southwest of the country once quality differences are taken into account. Nevertheless, local PR production remains promising for the central regions, where the need for soil-fertility conservation and restoration is highest.

Appleton (2001) summarizes these considerations: "Phosphate rock is a low value, high volume commodity with high transport costs, so the economic potential of a phosphate rock deposit will be determined to a large extent by its location in relation to domestic and international markets. Most commercial phosphate rock deposits are located close to the coast and in countries with efficient deep-water port facilities. If the transport infrastructure in a country is poorly developed and especially if railways or slurry pipelines are not available to transport the phosphate rock, it may be more cost-effective to import high-analysis fertilizers such as DAP, rather than to develop local resources. Phosphate rock resources sited in geographically remote and/or unfavourable locations at great distances from markets or from efficient transport facilities are unlikely to be economic to utilize for international markets, whereas local or perhaps regional use may be an economically viable option. Conversely, for agricultural areas located at a great distance from the coast, especially in landlocked countries, it may be more cost-effective to develop local phosphate rock resources for use as direct application fertilizer rather than to import manufactured fertilizers. High transport costs may be of less importance if the phosphate rock can be converted into a high value manufactured fertilizer product, although the quantity and quality of the phosphate rock resources may prevent this."

Economic considerations and policies to support PR adoption

At the macro level, economic effects are usually measured through changes in private or public sector income, i.e. consumer and producer surplus, rent and government-budget effects. In addition to these efficiency-related goals, redistributional and environmental criteria are usually introduced. PR application generates extra employment in both mining and processing and in the infrastructure sector. Therefore, it can mitigate out-migration. These sectors need complementary investments by the government. For many governments, the aggregate impact on public finance (over and above direct project effects), employment, migration, and the balance of payments is an important consideration.

Intensification, new varieties, better fertilizers and reduced transport costs shift the agricultural supply curve to the right, causing output increases and/or price reductions. Welfare gains differ sharply when the good is a non-tradable (sorghum, millet) as opposed to a tradable (cotton) or a commodity subject to government price support. In the case of a non-tradable good, increased production results in decreases in price, and consumers are the main beneficiaries. With a tradable good, it is possible that no price change will occur. Therefore, farmers gain, resulting in increased land rents, while the gain to consumers is small or negligible. In the absence of a more comprehensive analysis, the following effects of PR application are of wider concern.

First, to the extent that PR investment affects food crops, it is likely to increase overall food security. With supporting services in place, higher average yields per hectare will raise total food availability, causing local prices to decrease under the closed-economy assumption. Such a development will alleviate at least in part the problem of access to food for the urban poor. In the event of a year with lower rainfall, the influence of PR on production will still be felt, and the lower output level is likely to exceed that in a normal year without the application of PR. Thus, sustained PR use can reduce output fluctuations and stabilize prices. To the extent that PR substitutes for imports of TSP, PR has a positive effect on the balance of payments. As the use of PR also raises agricultural production, food imports will not need to be so high.

Second, PR application is one of several ways of improving soil fertility improvement. It is a matter of cost-effectiveness as to which method is the preferred one in a particular location. The discussion about the best way to improve soil fertility should not distract attention from its role as a powerful instrument for triggering sustainable intensification, improved productivity and higher incomes. As Sissoko (1998) has shown, among 12 instruments for stimulating sustainable agriculture in Mali, "valorisation/bonification des terres" stands out as one of the most effective. Others are: output price support, lower transaction cost and fertilizer subsidies.

Third, more extensive analysis of policies to promote sustainable intensification in Mali (Kuyvenhoven *et al.*, 1998c) shows that making more sustainable production conditions available through research and extension enhances resource efficiency. However, without special incentives for adoption, adjustment costs may well outweigh marginal benefits. The impact of these policy incentives differs among farm types. Fertilizer subsidies encourage less soil-depleting technologies in cash crops (cotton), but at the expense of more soil-depleting cereal activities. In this sense, fertilizer subsidies are inefficient in terms of resource use and reduce local food surpluses. As a result, food prices will increase, and farmers will be stimulated to revert to less soil-depleting methods. However, the food security of the net buyers of food (urban households) will now decrease because of higher prices, causing a policy dilemma.

As the case studies illustrate (below), the environmental benefits of PR application can be considerable, justifying some form of public or donor support to foster higher and faster rates of adoption where this appears desirable. The economic considerations mentioned above enable the identification of policy measures that may enhance farmer adoption of PR. A distinction will be made between price policies and institutional developments that contribute to creating a so-called 'enabling environment' for adoption, and extension programmes that support farmers' knowledge and information on PR.

Pricing policy

Few LDCs have been able to implement and sustain crop and fertilizer pricing policies to create stability and conduciveness in the pricing environment and to provide incentives for the adoption of fertilizer using crop technologies. Price stability should be ensured in countries that choose to develop their PR resources. Price stability and profitability are necessary for the adoption of fertilizer use in general and P as a capital investment by farmers in particular. Incentive pricing is needed to encourage farmers to use adequate quantities of appropriate plant nutrients, thereby sustaining agricultural production through investment in soil-fertility maintenance. The pricing policy should be such that it maintains favourable cost-benefit ratios for farmers' cropping activities compared with their off-farm activities (Teboh, 1995; World Bank, 1994).

In this context, two reasons can be advanced for considering fertilizer subsidies. First, only large-scale production can achieve economies of scale. At current prices, demand will never increase to levels necessary for realizing low production prices, while the latter do not decrease by lack of demand. Subsidies can break this vicious circle. Moreover, it allows the innovators to apply a new, politically and technically desired product at affordable prices. A subsidy that bridges the gap between the current prices and the lower price at large-scale production is relatively easy to manage. Furthermore, it disappears automatically when PR production increases. The second reason for subsidizing PR is the interest of the community in preserving the fertility of its soils, something that exceeds the individual interest of the farmer.

Organizational policy

Efficient and adequate organizational arrangements are essential for supplying fertilizers and other inputs on time, in the right type and quantity, and at the right price. Inefficiencies in creating adequate organizations have also constrained growth in fertilizer use in many African countries. In addition to fertilizer distribution, the distribution of PR as P capital imposes extra burdens on the marketing and distribution system because additional product tonnage has to be transported from the mining site to the consumption areas. Managing such large quantities requires organization at all levels, namely: mining, processing, transport, storage and marketing.

Agricultural product markets

Enhancing soil fertility is supposed to lead to higher production. For the individual farmer, this is the main rationale for applying PR. However, some or all of the additional produce needs to be marketed. In many countries, markets for agricultural products are fragmented. In West Africa, the cotton market is the best organized one in terms of marketing and distribution, including that of fertilizers. On the other hand, the application of modern technology and inputs for traditional cereals often remains limited as farmers are often unable to sell surplus production at reasonable prices. The promotion of PR in the interests of conserving and restoring soil fertility might require governmental measures in order to ensure an outlet for traditional farm products. However, the trap of reinstalling the old marketing hoards should be avoided. One solution could be to establish minimum prices well below normal market levels but to guarantee each desired sale. If these marketing problems are not addressed, technological innovations will remain confined to promising cash crops such as cotton, rice, certain vegetables and products with an institutionalized demand for the agro-industry (e.g. peanuts).

Risk protection

Agriculture depends considerably on climate conditions. In particular, the risks of farming are high in arid and semi-arid zones. Many farmers are reluctant to invest in agriculture or to contract loans in the face of such uncertainty. Credit institutions are often hesitant about granting loans because they fear repayment problems. This hampers the introduction of those new technologies that require capital investments.

Some kind of risk insurance could help farmers and banks to mitigate this problem. While prohibitive costs exclude comprehensive coverage of income risk, limited insurance systems might be possible, e.g. rescheduling of payback conditions in the event of crop failure, and temporary payment of interest out of a special fund. Government interventions to support farmers and credit institutions might be desirable in order to maintain financing risks at acceptable levels.

Research and extension policy

Despite considerable research and attention directed to the issues of technological adoption, a consensus has not developed regarding the social and economic conditions that lead farmers to adopt new production practices. It is unclear why some farmers adopt new technologies and others do not. As shown by Rauniyar and Goode (1992), adoption of technological practices tends to be interrelated. Therefore, programmes should emphasize the technological adoption of packages of practices rather than individual practices or a package containing all practices.

Countries in areas such as the Sahel still lack the necessary research and extension facilities for promoting fertilizer use and associated technologies. Research is needed to develop site-specific fertilizer recommendations, including the use of PR amendments, and to enable extension services to educate farmers and producers on fertilizer use and improve services. Because PRs differ in their reactivity levels and crops differ in their P requirements (demand and pattern), it is necessary to identify appropriate PR application rates for different cropping patterns in various agroclimatic zones. Applied agricultural research and extension services should be equipped to perform their different roles effectively. In addition to facilitating farmer application of recommended, research-based agricultural practices, extension services should assist farmers in understanding the environmental damage associated with various farming practices. Specifically, the rural poor should be made aware of the full implications of using technology provided by outside institutions (Teboh, 1995; World Bank, 1994).

The market and infrastructure perspective conceptualizes the diffusion of innovations as a process involving three principal activities: (i) establishment of a diffusion agency to make the innovation available to the client system; (ii) selection and implementation of strategies that include pricing and promotional communication in order to induce adoption; and (iii) adoption of the innovation by the client system. This perspective is complementary to the traditional adoption perspective. Farmers' resource endowments and attitudes, as well as the socio-political and physical environments in which they carry out their daily activities, are potential determinants of their adoption behaviour.

In general, farmers adopt innovations most rapidly when: (i) the farmers perceive that the innovations have a relative advantage over the practices they supersede; (ii) the innovations are compatible or consistent with their own values; (iii) the innovations are easy to understand and use; (iv) the innovations are amenable for experimentation on a limited basis; and (v) the innovations are capable of producing visible results (Rogers, 1983).

CASE STUDIES OF PR EXPLOITATION AND USE

The International Fertilizer Development Center (IFDC) and the World Bank conducted various case studies of PR use in the second half of the 1990s following international initiatives. Table 30 presents a summary of findings for Burkina Faso, Madagascar and Zimbabwe from a study by the World Bank (1997). Table 31 present findings for Mali from a study by the IFDC, based on the same methodology (for different regions and crops) (Henao and Baanante, 1999). Both studies report substantial rates of return at the assumed PR cost, important environmental benefits, but also better results at the farm level if alternative sources of industrial P fertilizer are applied.

Constraints of the type discussed above are recognized as limiting potential demand. Their removal, together with a strong political commitment and a comprehensive approach to PR exploitation, would be necessary in order to make PR application a successful proposition.

In the IFDC Mali study, risk and credit constraints are emphasized as factors that reduce the incentive to apply fertilizer. Another Mali study based on World Bank data (Kuyvenhoven *et al.*, 1998a, 1998b) analyses the income position of farmers adopting PR with loan financing, under the assumption of limited rainfall that reduces yields by 50 percent every third year. Under such circumstances, farmers would be unable to meet their financial obligations in bad years, and would need refinancing of their loans. This factor would limit both farmers' willingness to apply PR or other alternatives and the willingness of financial institutions to extend credit.

In a study of Burkina Faso, Hien *et al.* (1997) found that farmers faced with risk may forego the higher returns of PR over commercial fertilizer and choose the latter, more soluble product. Only under a limited supply of commercial fertilizer would farmers prefer PR, especially of the partially acidulated type.

In a review of fertilizer use in semi-arid West Africa, Shapiro and Sanders (1998) concluded that under existing farming conditions "imported inorganic fertilizers are the only technically efficient and economically profitable way to overcome prevailing soil-fertility constraints and alternative soil-fertility measures, such as organic fertilizers and natural rock phosphate, should be seen as complements to rather than substitutes for imported inorganic fertilizers, until wider experimentation makes them more successful."

As most case studies illustrate, PR application appears interesting and beneficial as it improves soil quality and increases yields. However, a PR project would also encounter numerous problems. Many farmers in LDCs lack complementary inputs and, therefore, are not able to apply PR adequately. During peak season periods, labour is scarce. Applying PR would increase labour demand particularly during these periods. With labour in short supply, a situation might occur where the farmer would not be able to profit from the benefits of the PR application because of delayed planting/seeding or the inability to harvest the increased yields fully. Other problems arise with the application of PR. PR development is a lumpy investment. As farmers cannot obtain credit from a bank because of poor collateral, they cannot finance PR purchases easily. Another problem concerns the willingness of farmers to adopt PR application as a technique for improving soil fertility. Research has shown that fertilizer use in low-income

TABLE 30
Matrix of costs and benefits associated with PR investments

Description	Burkina Faso	Madagascar	Zimbabwe
PR reserves			
Quantity (million tonnes)	63.0	0.6	47.4
Type	Sedimentary	Guano	Igneous
Reactivity	Moderate	High	Low
Incremental cumulative yield (kg/ha)[1]			
Maize	3 262	25 000	4 050
Sorghum	3 249		
Millet	1 801		
Paddy rice		5 020	
PR costs			
Ex-factory (US$/tonne)	152.6	103.4	37.3
Farm level (US$/ha)	78-110	92-372	35-45
PR benefits (NPV, US$/ha)[2]			
Private - total	131-137	3 878-7 492	500-517
Yield effects	122-128	3 878-7 492	500-517
BNF	9	854-938	
Environmental - total	317		210
Prevention of land degradation	208		147
Carbon sequestration	99		
Prevention of phosphogyp. pollution	10		63
Others (not quantified)			
Food security	High	High	Medium
Forex. savings	Medium	High	Low
Intergeneration equity	High	Medium	Medium
Impact on women farmers	Positive	Unknown	Unknown
Prev. of soil erosion/sedimentation	Medium	Medium	Low
Preservation of biodiversity/forests	Low	High	Medium
Other adverse environmental impacts overall:	Minimal	Manageable	Minimal
Ecological damage	Minimal	Moderate	Minimal
Land reclamation cost (US$/tonne)	Insignificant	2.2	1.0
Eutrophication	Minimal	Minimal	Minimal
Sharing of benefits (NPV, US$/ha) (%)[3]			
Private/local	270 (46)	2 775 (43)	867 (81)
National	208 (35)	3 534 (55)	147 (14)
Global	109 (19)	129 (2)	63 (5)
Description	Burkina Faso	Madagascar	Zimbabwe
NPV of alternate products (US$/ha)[4]			
Private			
Imported TSP	360		
DSP			1 302-1 313
Environmental			
Imported TSP	307		
DSP			216
Sensitivity analysis (NPV, US$/ha)[5]			
Private benefits	90-219		381-731
Total benefits	482-725	8 561-6 750	906-1 299
Potential demand (tonnes) P_2O_5 (PR)/year	21 400 (79 300)	21 300 (106 500)	20 500 (58 600)[6]
Constraints on PR investments			
Land tenure	Possible	Possible	Possible
Credit	Severe	Severe	Severe
Ecological (PR production)	None	Uncertain	None
Extension	Moderate	Moderate	Low
Competition from existing industry	None	None	High
Pricing environment	Unlikely	Unlikely	Unlikely
Conditions necessary for success			
Political commitment	Essential & available	Essential & available	Essential & unknown
Removal of constraints	Necessary	Necessary	Necessary
Investment in comp. packages	Essential	Essential	Essential

[1] From one-time basal application of PR.
[2] Private and environmental benefits are discounted by 10 and 3%, respectively.
[3] Based on 3% discount rate.
[4] Most profitable alternate product is included here.
[5] Discount rates range between 5 and 15% for private benefits and between 1 and 5% for total benefits.
[6] Phosphate rock concentrate.
Source: World Bank, 1997.

TABLE 31
Distribution of benefits for selected fertilizer strategies

Region	Cropping system	Fertilizer		Benefits*			
		Basal	Annual	Private	Environmental	Total	Environmental
		(kg/ha)		(US$/ha)			%
Sikasso	MC	TPR:120		647.8	403.5	1 051.3	38.4
		TPR:120	TSP:15	1 021.3	403.5	1 414.9	28.5
	MM	TPR:120		325.6	344.2	669.8	51.4
		TPR:120	TSP:15	633.5	344.2	967.8	35.6
	CC	TPR:120		600.1	412.8	1 012.9	40.8
		TPR:120	TSP:15	1 200.3	412.8	1 603.2	25.7
Segou	MiG	TPR:120		605.7	351.0	956.7	36.7
		TPR:120	TSP:15	793.7	351.0	1 134.8	30.9
	MiMi	TPR:120		380.4	345.2	725.6	47.6
		TPR:120	TSP:15	599.8	345.2	925.1	37.3
Kayes	GMa	TPR:120		390.3	377.8	768.1	49.2
		TPR:120	TSP:15	672.9	377.8	1 040.8	36.3
	GG	TPR:120		386.1	399.0	785.1	50.8
		TPR:120	TSP:15	672.9	377.8	1 040.8	36.3
Koulikoro	SG	TPR:120		788.6	370.9	1 159.5	32.0
		TPR:120	TSP:15	1 032.3	370.9	1 393.4	26.6
	SS	TPR:120		582.7	365.6	948.3	38.6
		TPR:120	TSP:15	728.9	365.3	1 084.6	33.7
Mopti	Rice	TPR:120		836.9	433.0	1 269 9	34.1
		TPR:120	TSP:15	1 111.9	433.0	1 545.0	28.0
	Millet	TPR:120		311.8	356.2	667.9	53.3
		TPR:120	TSP:15	499.8	356.2	846.1	42.1

* Discount rate: private benefits = 10 percent; environmental benefits = 3 percent.
Source: Henao and Baanante, 1999.

countries is extremely limited. Most knowledge of soil-improving techniques for food crops has been disseminated by the farmers themselves rather than by extension workers. Finally, lack of infrastructure may prevent timely and adequate supply of the PR to the farmers, which is one of the necessary conditions for convincing them to adopt this technology.

Before a PR application investment project can be introduced successfully, other measures and activities are needed. These include: (i) addressing credit problems and land rights; (ii) improving and extending rural infrastructure; (iii) improving marketing and distribution networks; and (iv) increasing the effectiveness of extension services. The securing of land rights for farmers will increase their ability to apply for credit and their willingness to invest in their land. Rural infrastructure needs to be improved and expanded in order to guarantee the timely and adequate supply of PR to the farmers. An improved rural infrastructure will also have a favourable impact on the distribution of produce to urban areas. Both farmers and the urban poor will then profit from reduced transport costs. Effective contacts between farmers and extension workers need to be developed so that extension services can be delivered effectively.

MARKETING OF INDIGENOUS PHOSPHATE ROCKS FOR DIRECT APPLICATION

Several countries have imported large quantities of PR for direct application. In 2000, Malaysia imported 500 000 tonnes, Brazil 320 000 tonnes, New Zealand 130 000 tonnes, and Indonesia imported about 230 000 tonnes in 1998. In other countries, indigenous PRs have been milled and marketed for domestic use on a relatively small scale with mixed results. PR sales depend on the supply of and demand for PR, competition with the imported P fertilizers, government policy, etc. The following sections provide information on the marketing of indigenous PRs in some countries where farmers have used PRs for crop production.

Latin America

In Colombia, there are several PR deposits. The only PR source being marketed for agronomic use is a medium-reactive Huila PR (product name: Fosforita Huila). In 1994, a company called Fosfacol sold 25 000 tonnes of Huila PR, about 15 percent of Colombia's annual consumption of P fertilizers. According to the company, sales are growing. However, no recent sales figures are available.

In Peru, there is a huge deposit of a highly reactive PR (Sechura or Bayovar). However, owing to various problems, there has been no large-scale milling of this deposit for marketing. Small quantities of Sechura PR are reportedly being marketed in the country and for export to Chile for direct application.

In Chile, the highly reactive Bahia Inglesa PR has been marketed for pastures and crop rotations in the volcanic-ash-derived soils in the south of the country. A company called Bifox markets the product under the name Fosfato Natural Chileno. Over the past 10 years, the average annual PR consumption has reportedly been about 10 000 tonnes of Fosfato Natural Chileno.

In Brazil, attempts to use the numerous PR deposits for direct application yielded poor agronomic results because of the low reactivity of the PRs, and direct application was discontinued. Some local PR sources are used to produce thermophosphates by fusion of mixtures of PR and basic slags at high temperatures.

In Venezuela, existing PR resources are not very reactive, except for one PR. A company called Pequiven is using this Riecito PR to produce commercial-grade PAPR products. The company markets the products under the name Fosfopoder and sells up to 150 000 tonnes/year.

Asia

In India, several PR deposits have been mined on a large scale for use in the chemical acidulation process. In the early 1980s, a company called Phosphate, Pyrite and Chemicals Ltd. began to promote a low-reactive Mussoorie PR (product name: Mussoorie Phos) for direct application. In 1993–94, sales of Mussoorie PR for agronomic use were about 107 000 tonnes. However, the supply of Mussoorie PR is reportedly declining because of a limited PR reserve and a new government policy to curtail PR mining. A company called Rajasthan State Mines and Minerals Ltd. has marketed a low-reactive Jhamarkotra PR (product name: Raji Phos) for plantation crops in acid soils in coastal areas of the southwest region.

In Sri Lanka, mining of a low-reactive Eppawala PR commenced in 1974 with a production level of 3 000–5 000 tonnes. At that time, the country was importing about 50 000 tonnes/year of PR for direct application. Since 1992, a company called Lanka Phosphate Ltd. has increased the production of Eppawala PR steadily to 32 000 tonnes/year while PR imports have decreased to 8 000 tonnes. In 1998, Eppawala PR consumption amounted to 13 000 tonnes of P_2O_5, representing about 43 percent of total P_2O_5 use in the country. Eppawala PR is used mainly for plantation crops, e.g. tea, mature rubber and coconut.

Sub-Saharan Africa

In terms of nutrients, P has been a major limiting factor on crop production in sub-Saharan Africa. The lack of a developed domestic P fertilizer industry and the limited availability of foreign exchange for fertilizer imports have prevented the resource-poor farmers from using expensive fertilizers. The average use of P fertilizers in sub-Saharan Africa is about 1.5 kg of P_2O_5 per hectare (in Asia and Latin America, the average figures are 34 and 20 kg of P_2O_5 per hectare, respectively). Numerous studies have attempted to determine whether indigenous PR deposits in sub-Saharan Africa can serve as sources of P in order to improve soil fertility and crop production.

In Mali, a medium-reactive Tilemsi Valley PR has been found suitable for direct application to acid soils for cotton, maize, rice, millet and sorghum. The entire production of PR is used within the country. The estimated production of ground PR was 4 529 tonnes in 1981, 8 092 tonnes in 1988, 11 000 tonnes in 1989, and 18 560 tonnes in 1990. Between 1991 and 1994, the production plant was often closed as a result of political unrest in the mine area. The estimated production potential is 36 000 tonnes/year.

In Burkina Faso, low-to-medium-reactive Kodjari PR has been mined on a small scale (about 1 000 tonnes/year on average). In 1993-94, PR production for direct application was reportedly 2 200 tonnes. The ground PR was promoted for a variety of crops including maize, rice and sugar cane. The estimated potential production capacity of Kodjari PR is 3 600–10 000 tonnes/year.

In Nigeria, a large reserve of PR of low-to-medium reactivity has been discovered recently in Sokoto State in the north of the country. Recent studies have shown that the agronomic effectiveness of Sokoto PR in terms of SSP was 40 percent on Alfisols, 100 percent on Ultisols, and 160 percent on Oxisols. The results of the field trials indicated that in the first, second and third years of cropping the agronomic effectiveness of Sokoto PR (in terms of SSP) was 54, 83 and 107 percent, respectively. A recently developed P product (product name: Crystal Super), produced by mixing Sokoto PR and natural talc magnesite ore, has reportedly been marketed by a company called Crystal Talc Nigeria, Ltd., in Kaduna. The company markets about 5 000 tonnes/year of this fertilizer.

In the United Republic of Tanzania, a highly reactive Minjingu PR deposit was developed in 1983. Production was about 20 000 tonnes/year (for acidulation) until the plant closed in the early 1990s. Of a recorded production of 2 500 tonnes in 1994, about 1 800–2 000 tonnes was exported to Kenya, where it was used for direct application and for SSP production, while 500–700 tonnes was sold for direct application in the north of the United Republic of Tanzania. The mine and processing plant now reportedly operate on a very limited basis.

Recently, the International Centre for Research in Agroforestry has reported promising results from agronomic trials on the use of Minjingu PR and a low-reactive Busumbu PR (Uganda) for different cropping systems, including agroforestry.

ECONOMIC CRITERIA IN PR ADOPTION AND USE – A CASE STUDY ON VENEZUELA

Venezuela plans to bring 46 percent more land under cultivation by 2018. The success of its efforts to expand the country's agro-industrial economy will depend on many factors. The most important of these factors relate to the large amounts of fertilizer required in order to achieve high agricultural productivity in tropical and subtropical areas with: (i) permanent crops, such as sugar cane, coffee, cocoa, fruits, cassava, improved pastures and forestry, covering more than 6 million ha; and (ii) annual crops, such as maize, sorghum, rice, leguminous grains, oilseeds and cotton on more than 700 000 ha. The prices of traditional water-soluble phosphate fertilizers, such as SSP, TSP, NPK fertilizers, mono-ammonium phosphate (MAP) and DAP, are high.

About 70 percent of the agricultural soils of Venezuela are acid, with low-activity clays and high P-fixing capacity. This situation, together with large reserves of PR and the increase in the cost of imported P fertilizers, has promoted a diversification in the production of P fertilizers, especially in fertilizer produced from PR sources in the country (Casanova *et al.*, 1998). A more rational use of P fertilizers has been proposed in order to counter dependency on imported fertilizers (Casanova, 1993).

Agronomic evaluation and domestic market potential

TABLE 32
Partially acidulated phosphate rocks based on various Venezuelan deposits

Phosphate rock	Partially acidulated phosphate rocks	
	Concentration (% P_2O_5)	Acidulation grade (%)
Riecito	25-33	40-60
Monte Fresco	25-34	15-40
Navay	20-24	40

The search for the most efficient ways of using the country's indigenous phosphate resources began with the sampling of three major existing PR deposits in Venezuela: Monte Fresco and Navay (Tachira State), and Riecito (Falcon State). A pilot plant produced PAPRs with H_2SO_4 and H_3PO_4 acids. Table 32 shows the main characteristics of the products.

Studies assessed the agronomic efficiency of the natural PRs and these PAPRs on various crops and soil types. The investigations focused on finding a process route for the production of low-cost phosphate fertilizers from the partial acidulation of indigenous PRs. Various national research institutes, universities, technical assistance companies and producers took part. International organizations, such as the IFDC (the United States of America), the Centre de Cooperation Internationale en Recherche Agronomique pour le Developpement (CIRAD) and Technifert (France), CIAT (Colombia) and FAO/IAEA, provided additional assistance. The agronomic efficiencies of these PAPRs (20–30 percent P_2O_5 concentrations and acidulation grades of 40 percent and above) using conventional and isotopic techniques were assessed on various crops and soil types. They yielded excellent results in comparison with traditional fertilizers. Compared with the national average, yield increases of 29–100 percent were obtained in sorghum, maize, sugar cane, soybean and pasture (Casanova, 1998; Casanova *et al.*, 1998).

The results of the above evaluations, some of which spanned a period of more than 10 years, led to the following conclusions: (i) agronomic yields and efficiencies comparable to those with TSP were achieved by using PAPRs with 20–30 percent P_2O_5 concentrations and acidulation grades of 40 percent and above on annual crops such as sorghum, soybean, maize and permanent crops such as pasture and coffee; and (ii) a medium-term forecast of the potential domestic market for PAPRs of some 490 000 tonnes/year.

Economics of PAPR production

The next step was to establish the most economic route to convert PAPR profitably. The following factors were taken into account: (i) PR extractable and economically viable deposits, mining infrastructure and capacity of installed production; and (ii) PAPR production process (process method, other raw materials, existing process synergy and product quality).

PR source

PR from the Riecito deposit proved to be the most attractive option as the deposit was already under commercial extraction by a company called Pequiven. There are both sufficient reserves and sufficient installed production capacity at the Riecito mine to support a production project that could satisfy the estimated potential market for PAPR in Venezuela.

The production capacity for the initial project was set at 150 000 tonnes/year of PAPR, equivalent to one-third of the estimated potential domestic market for PAPR. The Riecito deposit can satisfy more than 20 years of PAPR production at this capacity, in addition to meeting the rock feedstock requirements of other phosphate production plants (DAP and NPKs) at the Moron Complex. Table 33 shows the current installed P_2O_5 capacity at the complex.

TABLE 33
Current production capacity for P_2O_5 at the Moron Complex

Product	Tonnes/year	P_2O_5 (tonnes/year)
PAPR	150 000	37 500
H_3PO_4	200 000	90 000
Total	350 000	127 500

Production process

From the process point of view, it is possible to produce PAPR commercially in several ways along the same lines as SSP and TSP. The preferred process should be one that gives the highest possible return on investment. With this aim in mind, the following factors were taken into account: the quality of the Riecito PR, process simplicity, integration with existing facilities, and the availability of other raw materials and services.

PR quality was a very important factor in deciding which process route to use. Based on the expertise accumulated over seven years of assessment of Riecito PR (acidulation, digestion, granulation, etc.) at the Petroleos de Venezuela Research and Technological Support Centre and a survey of the available technologies, it was determined that the most attractive process for Pequiven's specific conditions should have the specific technical features. These features are: (i) it should have the facilities for combined acidulation with sulphuric and phosphoric acids in order to provide the flexibility to produce PAPR ranging from 20 to 30 percent P_2O_5; (ii) it should be capable of using diluted H_3PO_4 with a wide range of solids content (26 percent P_2O_5 and 0–30 percent solids) in order to allow efficient use of clarifier underflow from the Moron phosphoric acid plant; and (iii) it should perform the two operations of acidulation and granulation of the PAPR in a single stage for the sake of process simplicity.

The chosen process for the Moron PAPR plant is based on the one-step PR-sulphuric acidulation method developed by the IFDC (Schultz, 1986). However, with the accumulated expertise of the Petroleos de Venezuela Research and Technological Support Centre on Riecito PR acidulation, it was possible to simplify the process by removing the cooling stage for the PAPR product. Figure 32 shows how the acid for PAPR production is taken from the phosphoric acid plant. The PR digestion stage in the dehydrate reactor has been optimized for the reduced solids content of the phosphoric acid recycled to the reactor, resulting from the withdrawal of

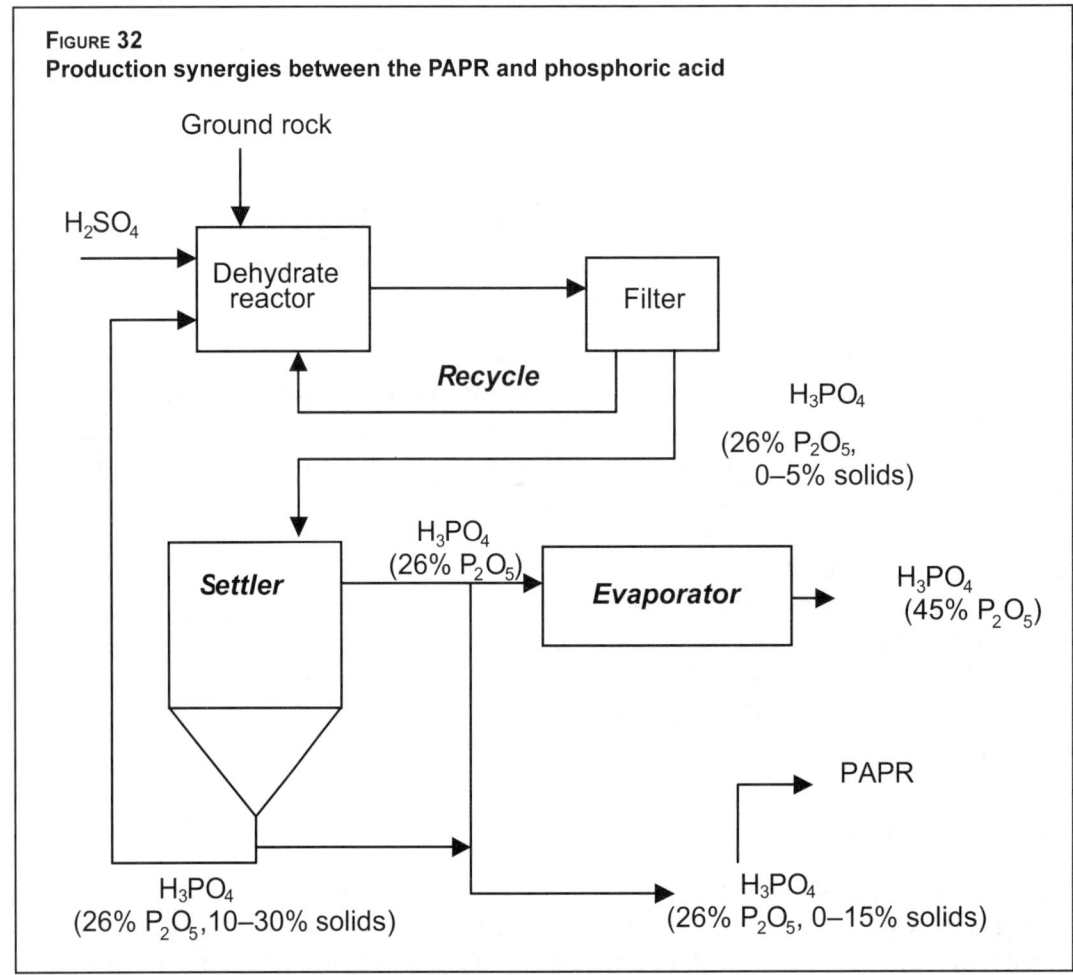

FIGURE 32
Production synergies between the PAPR and phosphoric acid

the filter-strength acid and clarifier bottoms for PAPR prior to entering the evaporation stage. It is estimated that this increase may represent up to 10 percent of the production capacity in the phosphoric plant. This has a significant impact on the economics of phosphate fertilizer production (DAP, NPK, PAPR) at the Moron Complex and augurs well for the profitability of the investment in PAPR production at the site.

Pilot-plant evaluation test

A pilot plant (500 kg/h of PAPR) evaluation was set up to test the proposed production process at the IFDC's facilities in the United States of America (IFDC, 1996). The evaluation results confirmed the feasibility of producing PAPR via the one-step acidulation/granulation process, using a combination of acidulants including H_2SO_4, H_3PO_4 and diluted H_3PO_4 with a wide range of solids content (26 percent P_2O_5, 0–30 percent solids). It also confirmed that the use of a product cooler was not necessary, as the resulting Riecito PAPR had excellent physical properties. Table 34 provides a summary of the results obtained during the pilot-plant assessment. These results helped develop the design basis for a commercial plant of 150 000 tonnes/year of PAPR at the Moron Complex.

TABLE 34
Pilot-plant evaluation results

Item	Results	Comments
One-step granulation/acidulation process	Feasible	Steady process
Combined use: H_3PO_4 (26% P_2O_5, 0-30% solids) & H_2SO_4 (90-98%)	Feasible	
Granulation efficiency	> 90%	
Acidulation grade	40-60%	Predictable/calculable
PAPR concentration (% P_2O_5)	25-30%	Calculable
Product quality		
CRH (at 30 °C)	85-90%	High (excellent) CRH
Hardness	> 4 kg	Typical fertilizer > 2 kg
Impact resistant	< 1.5%	Typical MAP/DAP < 1.5%
Abrasion resistant	< 1.5%	Typical MAP/DAP < 2%
Hygroscopicity (at 30 °C) (mg/cm³)	< 60	No hygroscopic typical fertilizer < 150

PAPR quality

Taking into account the agronomic assessment of various PAPRs (from Riecito and other sources) and the desirability of having a PAPR with a P_2O_5 grade close to the average P_2O_5 grade of the Riecito PR *in situ*, it was decided that the PAPR produced at the Moron Complex should have a P_2O_5 concentration of 25 percent and an acidulation grade of 50 percent. However, it was also considered desirable that the plant should have the flexibility to produce PAPR in the range of 20–30 percent P_2O_5 concentration and an acidulation grade of 40–60 percent.

In order to establish and forecast the required proportions of sulphuric acid, phosphoric acid and PR that may be necessary to produce a certain PAPR quality (P_2O_5 content, acidulation grade), a calculation programme was developed based on the reaction stoichiometry of Riecito PR in combined acidulation. The programme predicts the optimum relationship between the acidifiers needed and enables the process conditions and the associated variable costs to be established.

The Moron PAPR plant

The design project for the PAPR plant commenced in 1995, with the detailed engineering, equipment procurement and construction stage beginning in 1997. The production capacity of 150 000 tonnes/year (19 tonnes/h, on a basis of 24 h/day with a service factor of 330 d/year) determined the process equipment sizes. Since the completion of the installation guarantee test in November 1998, the PAPR produced has achieved both the physical and chemical quality standards required. The PAPR fertilizer is marketed in Venezuela under the name Fosfopoder.

Figure 33 shows the process flow chart with the six sections that make up the plant. The PAPR plant has a distributed control system (whereby the plants receives automatic information). The operation panels of this system are located in the phosphate control area, as are all the Moron Complex's operation control functions (Dominguez and Barreiro, 1998).

Raw materials and products

The PAPR plant at Moron has the flexibility to use a wide range of concentrations of raw materials. The expertise of the operations staff and the calculation programme installed in the control panel enable the product quality and production to be adjusted (Castillo *et al.*, 1998).

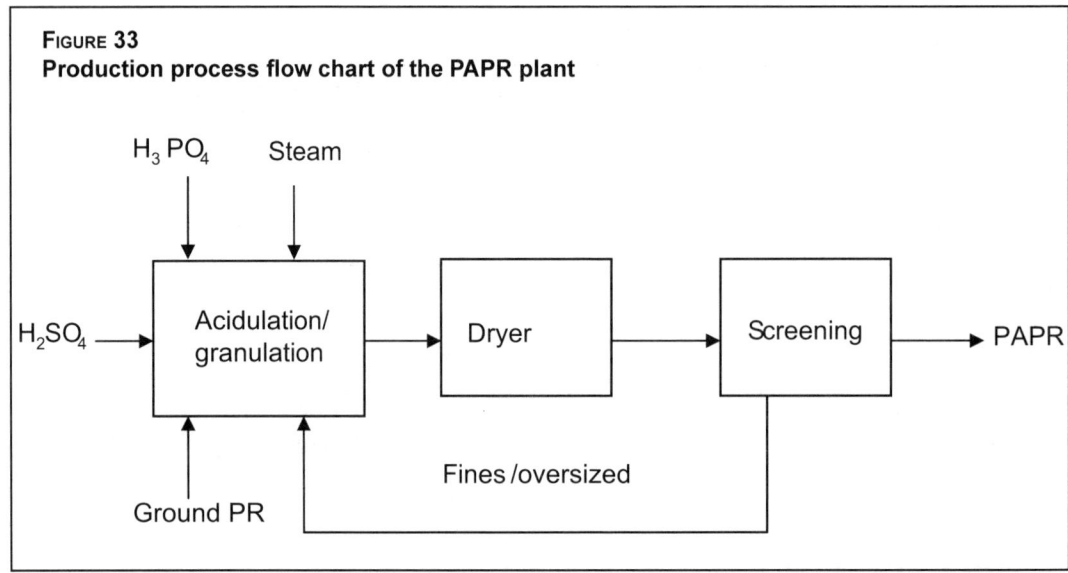

FIGURE 33
Production process flow chart of the PAPR plant

The calculation programme is based on the chemical equations and the material balance for the acidulation of Riecito PR with sulphuric and phosphoric acid and the heat transfer equations occurring in the process. In this way, it is possible to estimate the theoretical proportions of raw materials to product. The quality of the PAPR produced (grade, and acidulation percentage) has validated these calculations.

With regard to product quality, the use of 40–60-percent acidulated PR can achieve a PAPR quality with chemical specifications and physical qualities similar to other PAPRs with an acidulation of 40–60 percent.

In 1999, the first full year of operations, sales to the domestic market were about 15 000 tonnes of PAPR (10 percent of plant capacity). This can be considered something of a success given that PAPR is a new product for the Venezuelan market. In the near future, the plant will focus on the production of 25-percent P_2O_5 and 50-percent acidulation PAPR. The production is marketed under the generic brand name of Fosfopoder.

Chapter 11
Legislation and quality control of phosphate rocks for direct application

Many countries, especially developing countries, have legislation to ensure that the physical and chemical qualities of commercial fertilizers meet the specifications set by the governments in order to safeguard consumers' interests. However, in some developed countries, e.g. the United States of America, it is possible to sell fertilizers without specifications provided they meet the 'truth in labelling' requirement.

In 1998, the consumption of direct application phosphate rock (DAPR) represented less than 2 percent of world P_2O_5 consumption (Maene, 2003). The information available in literature on the legislation for phosphate rock (PR) as a fertilizer for direct application is similarly limited. This chapter reviews some of the existing legislation in various countries concerning DAPR. It then discusses issues associated with the legislation and recommends legislation guidelines for the quality control of DAPR.

CURRENT LEGISLATION ON PHOSPHATE ROCK FOR DIRECT APPLICATION

Most legislation concerning PR for direct application includes three main specifications: total P_2O_5 content of PR, chemical property (solubility), and physical property (particle size). The following sections present examples of legislation adopted in countries that use DAPR.

In 1976, the Council of the European Communities (EC) issued a directive (Official Journal of the European Communities, 1976) regarding the specifications of DAPR as follows: (i) minimum 25 percent total P_2O_5; (ii) at least 55 percent of total P_2O_5 soluble in 2-percent formic acid (FA); and (iii) at least 90 percent through 0.063 mm (250 Tyler mesh) and 99 percent through 0.125 mm (115 Tyler mesh) in particle-size distribution.

In Malaysia, PR has been in widespread use since the 1950s for plantation crops (e.g. oil-palm and rubber). In 2000, Malaysia imported almost 500 000 tonnes of various PR sources. In 1998, the Department of Standards of Malaysia revised the previous 1972 specifications as follows: (i) minimum 28 percent total P_2O_5; (ii) at least 7.5 percent P_2O_5 (weight basis) soluble in 2-percent citric acid (CA); and (iii) 90 percent less than 0.500 mm (32 mesh) in particle-size distribution (Malaysian Standard, 1998). Thus, the unground, as-received PRs can now meet the particle-size requirement stipulated by legislation in Malaysia.

In Brazil, the use of imported, unground, as-received, highly reactive PRs, e.g. Gafsa (Tunisia), Arad (Israel), Daoui (Morocco) and Djebel Onk (Algeria), increased sharply to about 320 000 tonnes in 2000 (ANDA, 2001). The Brazilian Ministry of Agriculture publishes the regulations for "as-received, natural reactive phosphate" (Regulations Nos. 9, 63, 161 and 19). These specify: (i) minimum 28 percent of total P_2O_5; (ii) at least 9 percent P_2O_5 (weight basis) soluble in 2-percent CA; (iii) 100 percent less than 4 mesh (4.8 mm) and 80 percent less than 7 mesh (2.8 mm) with a tolerance of 15 percent for particles larger than 4.8 mm; and (iv) calcium (Ca) content of 30–35 percent.

In India, the consumption of PRs, mainly indigenous low reactive and low grade for direct application, was about 25 000 tonnes as P_2O_5 in 1998 (Maene, 2003). Imports of PR are subject to an import duty of 5 percent. The Indian Fertilizer (Control) Order issued in 1985 for PR is: (i) minimum 18 percent of total P_2O_5; and (ii) 90 percent less than 100 mesh (0.15 mm) and the remaining 10 percent must be less than 60 mesh (0.25 mm). There is no minimum solubility requirement of PR in the legislation.

In New Zealand, the use of imported, reactive PRs for direct application increased from 102 000 tonnes in 1996 to 130 000 tonnes in 2000, about 10 percent of all phosphorus (P) products sold on both a tonnage and dollar basis (Quin and Scott, 2003). In New Zealand and Australia, there is no legislation on DAPR. In general, reactive PR is defined as 30 percent of total P_2O_5 being soluble in 2-percent CA under standard conditions (Quin and Scott, 2003; Hedley and Bolan, 2003). Compared with 2-percent CA and neutral ammonium citrate (NAC), the 2-percent FA test best predicts the agronomic effectiveness of PRs varying in reactivity and particle size (Hedley and Bolan, 2003; Bolland and Gilkes, 1997). However, 2-percent CA remains the standard test in New Zealand and Australia.

ASSOCIATED ISSUES

The main factors affecting the agronomic effectiveness of DAPR: (i) inherent PR properties; (ii) type of soil; (iii) crop species; (iv) management practices; and (v) agroclimatic conditions (Chien and Menon, 1995b; Hammond *et al.*, 1986b; Khasawneh and Doll, 1978; Rajan *et al.*, 1996). Chapter 5 provides more details on these aspects.

All regulations concerning DAPR focus on the quality control of different PR sources, namely: (i) total P_2O_5 content; (ii) PR solubility; and (iii) particle-size distribution of PR. Thus, there is a need to discuss the issues of these physical and chemical properties of PR associated with legislation for DAPR.

Total P_2O_5 content of PR

TABLE 35
Total P_2O_5, NAC-soluble P_2O_5, and $CO_3:PO_4$ ratio of apatite in various phosphate rocks

PR source[a]	$CO_3:PO_4$	Total P_2O_5	NAC-P_2O_5[b]
		(% of rock)	
North Carolina, United States of America	0.26	30.0	7.6
Arad, Israel	0.20	32.4	7.1
El-Hassa, Jordan	0.16	31.3	5.8
Hahotoe, Togo	0.11	36.8	3.9
Idaho, United States of America	0.08	32.3	3.5
Kaiyang, China	0.05	17.6	3.4
Araxa, Brazil	< 0.01	36.1	2.8
Dorowa, Zimbabwe	< 0.01	33.1	1.9
Sukulu Hills, Uganda	< 0.01	41.0	1.6

a. Ground to minus 100-mesh (-0.15 mm) size.
b. NAC, second extraction.

The total P_2O_5 content of a PR bears no relationship to its chemical reactivity and agronomic effectiveness. In fact, a PR with a very high total P_2O_5 content (33–40 percent) may indicate that the PR is potentially low in chemical reactivity. As more CO_3 substitution for PO_4 in an apatite structure results in a lower total P_2O_5 content and higher chemical reactivity, a very high total P_2O_5 content suggests that the apatites in the PR have very low CO_3 substitution in the apatite structure and, hence, low chemical reactivity

(Rajan et al., 1996). Table 35 presents data on some PRs that vary widely in the $CO_3:PO_4$ ratio of apatite, total P_2O_5 content, and solubility in NAC solution. The solubility is related closely to the $CO_3:PO_4$ ratio of apatite but not to the total P_2O_5 content. Therefore, the inclusion of a minimum total P_2O_5 content in legislation should be based not on agronomic considerations but only on the consideration of protecting PR users from unscrupulous suppliers.

SOLUBILITY OF PHOSPHATE ROCK

The agronomic effectiveness of PR depends considerably on its chemical reactivity. This is determined conventionally by measuring the solubility of the PR in extracting solutions. The three most commonly used extraction methods (Chien and Hammond, 1978) are:

- NAC: A 1-g sample of PR is extracted with 100 ml of NAC solution at 65 °C for 1 h. This method is used mainly in the United States of America.

- 2-percent CA: A 1-g sample of PR is extracted with 100 ml of 2-percent CA at room temperature for 1 h. This is probably the most common method of the three in countries where PR has been used, e.g. Malaysia, Brazil and New Zealand.

- 2-percent FA: A 1-g sample of PR is extracted with 100 ml of 2-percent FA at room temperature for 1 h. This method is used in countries of the EC.

Although a given method specifies the extracting conditions such as solid-solution ratio, temperature, and time of extraction, there are no specifications on other conditions that may affect the solubility measurement, e.g. types of shaking apparatus (horizontal shaker, wrist-action shaker, reciprocal shaker, magnetic stirring bar, etc.), shaking or stirring speed during extraction, and size of shaking flasks used. Therefore, there is no universal method for measuring PR solubility.

However, even if a method did specify all the conditions for measuring PR solubility, there are other issues that would affect measurements.

Expressing PR solubility

The solubility of PR is commonly expressed as a percentage of total P_2O_5. For example, a PR has 30 percent total P_2O_5, and the NAC-soluble P_2O_5 is 6 percent when a 1-g sample is extracted with 100 ml of solution. The solubility of the PR is then expressed as 20 percent of total P_2O_5 in NAC. However, when comparing the solubility of PR sources that vary widely in total P_2O_5 content, the solubility is expressed more appropriately as a percentage of rock (weight basis) (Chien and Hammond, 1978; Chien, 1993, 1995). The expression 'percentage of total P_2O_5' may give a misleading comparison when PR sources containing low total P_2O_5 content are used. Table 36 shows this P_2O_5-grade effect. The NAC solubility expressed as a percentage of PR and sand mixture was relatively constant (5.5–6.0 percent) down through the 50:50 PR-sand mixture. On the other hand, citrate solubility expressed as a percentage of total P_2O_5 increased from 21.0 to 61.8 percent when total P_2O_5 of PR was diluted from 27.2 to 7.6 percent. The constant citrate solubility expressed as percentage of mixture was due to the fact that the solubility of a given apatite is fixed by its solubility-product constant (Chien and Black, 1976).

In a study to compare various laboratory methods for predicting the agronomic potential of DAPRs, Chien and Hammond (1978) observed that the solubility of two low-grade PRs (Pesca and Huila) appeared to increase with respect to other sources when the solubility was expressed

TABLE 36
Citrate solubility of mixtures of North Carolina PR and sand

Composition of mixture		Total P_2O_5 content	Neutral ammonium citrate solubility	
PR	Sand		of mixture	of total P_2O_5
		%		
0.91	0.09	27.2	5.7	21.0
0.86	0.14	25.7	5.9	23.0
0.80	0.20	23.9	6.0	25.1
0.69	0.31	20.6	5.6	27.2
0.65	0.35	19.4	5.7	29.4
0.50	0.50	14.9	5.5	36.9
0.25	0.75	7.6	4.0	61.8

Source: Lehr and McClellan, 1972.

TABLE 37
Citrate solubility of PRs containing various amounts of total P_2O_5 content

PR source	Total P_2O_5 content	Neutral ammonium citrate solubility	
		of rock	% of total P_2O_5
		%	
Huila, Colombia	20.9	3.5	16.2
Pesca, Colombia	19.8	1.8	9.5
Sechura, Peru	29.9	5.3	18.0
Gafsa, Tunisia	29.9	5.5	18.6
North Carolina, United States of America	30.0	6.7	22.4
Central Florida, United States of America	32.7	3.2	9.7
Tennessee, United States of America	30.1	2.8	8.9

Source: Chien and Hammond, 1978.

as a percentage of total P_2O_5 rather than as a percentage of rock (Table 37). For example, Huila PR and Central Florida PR had about the same citrate solubility expressed as a percentage of rock. When the solubility was expressed as percentage of total P_2O_5, Huila PR appeared to have a higher solubility than Central Florida PR. Braithwaite *et al.* (1989, 1990) also reported that the total P_2O_5 content of PR affected its solubility as measured by 2-percent CA and 2-percent FA; when a 1-g sample of any PR was used for assessment, PRs with a lower P_2O_5 content had an advantage over PRs with higher total P_2O_5 content.

Therefore, it appears that PR solubility expressed as percentage of rock rather than as percentage of total P_2O_5 provides a more accurate assessment when comparing the solubility of PR sources that vary widely in total P_2O_5 content.

Free carbonate effect

It is the apatite-bound carbonate and not the total carbonate (which may also contain free carbonates, such as calcite and dolomite, as accessory minerals in PR) that determines the solubility of PR. The presence of free carbonates in a significant amount can suppress the solubility of the PR. The apparent decrease in the solubility of apatite is due to the Ca-common-ion effect and consumption of extracting solutions, which occur because free carbonates are more soluble than apatite.

Chien and Hammond (1978) measured the reactivity of seven PRs by various chemical extractions (Table 38). The P solubility in NAC remained relatively constant between the first and second extractions for each PR except Huila PR. The Huila PR contains about 10 percent free $CaCO_3$ that suppressed the apatite solubility during the first extraction but not during the

TABLE 38
Solubility of PRs as measured by various chemical extractions

PR source	Soluble P_2O_5, % of rock			
	Neutral ammonium citrate		2% citric acid	2% formic acid
	1st extraction	2nd extraction		
Huila, Colombia	0.9	3.5	5.3	6.2
Pesca, Colombia	1.8	1.8	6.9	5.3
Sechura, Peru	5.3	5.3	15.2	21.9
Gafsa, Tunisia	4.8	5.5	14.0	22.3
North Carolina, United States of America	7.1	6.7	15.9	25.8
Central Florida, United States of America	3.0	3.2	8.5	8.3
Tennessee, United States of America	2.5	2.8	8.7	6.9

Source: Chien and Hammond, 1978.

second extraction. MacKay *et al.*, 1984 also found that the citrate-soluble P_2O_5 of Chatham Rise PR (New Zealand), which contains 27.6 percent free $CaCO_3$, increased from 3.2 percent in the first extraction to 10.1 percent in the second extraction.

The data in Table 38 indicate that the solubility of Huila PR was still suppressed by the free $CaCO_3$ in the PR as measured by 2-percent CA and 2-percent FA when comparing the solubility of Huila PR and Central Florida PR in the second extraction with NAC. However, the degree to which the free $CaCO_3$ affected the solubility of Huila PR appeared to decrease as the strength of the extracting solution increased from NAC to 2-percent CA and to 2-percent FA. To measure the actual solubility of a PR, a second extraction with NAC and 2-percent CA is therefore recommended when comparing the solubility of PR sources containing various amounts of free carbonates. For 2-percent FA, one extraction may suffice if the PR does not contain a significant amount of free $CaCO_3$.

Particle-size effect

The solubility of PR increases with decreasing particle size (Chien, 1993, 1995). However, a very fine grinding of a low-reactivity PR cannot increase its solubility significantly enough to compensate for the nature of its low reactivity owing to low CO_3 substitution for PO_4 in the apatite structure. Table 39 shows that the solubility of finely ground low-reactivity PRs is still lower than unground, as-received, high-reactivity PRs. The solubility of finely ground, highly reactive North Carolina PR was significantly higher than its unground form (Table 39). However, there were reportedly no significant differences in agronomic effectiveness obtained with finely

TABLE 39
Solubility of ground and unground PRs

PR source[a]	Soluble P_2O_5, % of rock	
	Neutral ammonium citrate	2% citric acid
Gafsa (Tunisia), unground	6.3	10.6
North Carolina (United States of America), unground	6.4	10.6
North Carolina (United States of America), ground	7.1	15.9
Hahotoe (Togo), ground	4.1	7.6
Kaiyang (China), ground	3.4	7.2
Araxa (Brazil), ground	2.8	5.0
Jhamarkota (India), ground	0.6	1.5

a. Unground = 95% < 32 mesh (0.50 mm); ground = 100% < 100 mesh (0.15 mm).

ground and unground forms for highly reactive PRs, e.g. North Carolina (Chien and Friesen, 1992) and Gafsa (Chien, 1998). Thus, it is not necessary to grind highly reactive PRs finely for direct application. Indeed, several highly reactive PRs for agronomic use are in unground, as-received forms, e.g. Gafsa PR (Tunisia), Djebel Onk PR (Algeria), and Sechura PR (Peru). The use of unground PRs for direct application saves on grinding cost and also reduces the dustiness of PR during handling and application.

As the agronomic effectiveness of PR depends on its solubility rather than particle size, the regulations regarding DAPR should not require all PRs, including the highly reactive types, to be ground to less than 100 mesh (0.15 mm). The minimum solubility requirement should be based on the PR products actually used, be they ground or unground.

GUIDELINES FOR LEGISLATION ON DAPR

Because PRs vary widely in chemical and mineralogical composition, reactivity, particle-size distribution, and accessory minerals, it is important to recognize that any regulations concerning a PR for direct application should be related to its agronomic effectiveness. For example, a highly reactive, as-received, unground PR can be more agronomically effective than a finely ground, low-reactivity PR. If a regulation requires that PR be ground to less than 100 mesh (0.15 mm), then farmers may not be able to use more reactive, unground PRs that may provide more agronomic and economic benefits than the use of finely ground, low-reactive PRs for crop production.

The following guidelines are proposed for consideration by regulatory agencies involved in devising legislation on DAPR:

1. As the total P_2O_5 of a PR is not related to its agronomic effectiveness, a regulation that requires minimum total P_2O_5 content should be solely for the purpose of protecting users' from unscrupulous suppliers.

2. All of the current methods of solubility measurement for PR require only one extraction. As associated free carbonates and particle size can affect the solubility of apatite, it is recommended that the sequential two-extraction procedure be carried out in order to obtain a more actual solubility value of PR in the second extraction. For example, if a PR has a small fraction of very fine particle size, the solubility obtained in the first extraction can be relatively high because of the small particle-size effect. During the second extraction, the solubility measured represents the more actual solubility of the main bulk of PR particles after small particles are dissolved in the first extraction. The two-extraction procedure can also eliminate the depressive effect of free carbonates on apatite solubility of PRs containing a moderate content of free carbonate (e.g. less than 10 percent). Thus, it is recommended that a sequential two-extraction procedure be used with NAC and 2-percent CA. For 2-percent FA, this may not be necessary because of its relative strength that may eliminate the effects of free carbonates and small particle size on apatite solubility during the first extraction.

3. The PR solubility value can be elevated artificially if it is expressed as a percentage of total P_2O_5. Therefore, when comparing PRs containing low total P_2O_5 content with PRs containing high P_2O_5 content, it is recommended that PR solubility be expressed as a percentage of rock. This eliminates the possible grade effect. This is particularly important when comparing solubility values of PRs that vary widely in terms of total P_2O_5 content.

4. For highly reactive PRs, the as-received, unground forms are almost as effective as the finely ground forms in agronomic use despite the fact that the solubility is lower with the unground forms. For low-to-medium-reactive PRs, both agronomic effectiveness and solubility are lower with the unground than the finely ground forms. Therefore, legislation should specify separate requirements for particle-size distribution for highly reactive and other less reactive PRs.

5. Although solubility affects the reactivity and potential agronomic use of the PR considerably, several other factors, such as soil pH and crop species, are also important in determining the ultimate agronomic effectiveness of the PR. For example, a highly reactive PR may not be effective for most of the food-grain crops grown on soils with a pH of more than six. A low-reactive PR may be agronomically effective for plantation crops grown on acid soils. Therefore, minimum solubility requirements should consider agronomic factors such as soil pH and crop species. One possible way of doing this is to establish a minimum solubility requirement of PR in different categories corresponding to different soil pH and crop species conditions.

6. The quality requirements for PR for chemical acidulation should not be the same as those for direct application. The factors that affect the acidulation process are not the same as those that are important for DAPR. For example, the sesquioxide ($Al_2O_3 + Fe_2O_3$) content of PR is critical for the production of H_3PO_4 and water-soluble P fertilizers whereas it is not a significant factor for DAPR. Therefore, legislation on quality control of DAPR should not be the same as that for acidulation.

Chapter 12
Epilogue

The appropriate and sound utilization of phosphate rocks (PRs) as sources of phosphorus (P) can contribute to worldwide development by fostering sustainable agricultural intensification, particularly in developing countries endowed with indigenous PR resources, besides minimizing pollution in countries where PRs are processed industrially. PR products are an agronomically and economically sound alternative P input to manufactured superphosphates and have the potential to remove the major constraint of low soil P in tropical and subtropical soils. As many developing countries in the tropical and subtropical regions import P fertilizers, the direct utilization of indigenous raw materials such as PRs would contribute not only to import substitution and energy and capital savings but also to fertility improvement of the impoverished soils and control of land degradation.

Although excellent world surveys of PR deposits and occurrences are available, a detailed re-assessment, particularly of those PR deposits that are most suitable for direct application, is highly desirable in view of the dynamic changes in the technology and economics of mining and processing as well as in their potential use in developing countries. The grade of direct application phosphate rock (DAPR) is not a critical factor from a technical standpoint. PRs with less than 20 percent P_2O_5 have been used on a local basis. However, grade (nutrient content) is an important economic consideration, particularly when evaluating transportation costs.

Advances in standard characterization, methods of evaluation and technologies for enhancing the agronomic effectiveness of PRs will help improve knowledge and management techniques for increasing adoption by farmers. With these scientific and technological advances, it would then be possible to identify specific management practices for effective and economic direct application of PRs with a view to fostering sustainable agricultural intensification in developing countries of the tropics and subtropics, while contributing to improved food security and controlling soil degradation. However, specific technologies need to be developed in each case and more research is needed in order to obtain conclusive results.

Technologies for the direct application of PRs have already been proved under a variety of agro-ecological conditions. However, their agronomic effectiveness depends ultimately on a wide range of factors and their interactions. Therefore, it is important to develop and pilot test decision-support systems (DSSs) in order to predict their effective and economic utilization. Further validation and refinement of these systems for more accurate prediction should be an ongoing process. The availability of a global phosphate rock decision-support system (PR-DSS) will be a useful research and extension tool for researchers, extension workers, progressive farmers, planners and agribusiness dealers. It will assist in promoting the use of PR resources in tropical and subtropical developing countries. It is also suggested that PR-DSS should be an integral part of a broader decision system for integrated nutrient management/soil productivity improvement or even of the Decision Support System for Agrotechnology Transfer (DSSAT).

A widespread promotion of DAPR would require: (i) standardized methods and procedures for their characterization and evaluation; (ii) reliable prediction of their agronomic effectiveness;

and (iii) predictability of crop yield increases and their profitability. Recent advances in scientific knowledge and technological developments on the utilization of PR need to reach all stakeholders involved in agricultural development. Conventional and advanced information technology, particularly participatory approaches, should be used to disseminate the DAPR technologies to farmers.

It is now known that effective PR sources provide not only P for plant growth but that they may also supply secondary nutrients, such as calcium and magnesium, and micronutrients, such as zinc and molybdenum, depending on the chemical and mineralogical composition of PR. Reactive PRs or PRs containing free carbonates (calcite and dolomite) can also raise soil pH to partially reduce aluminium (Al) saturation of acid soils and decrease Al toxicity to plant growth, although their liming effect is generally lower than lime. One environmental issue regarding PR applications has been the potential uptake by plants of elements that are harmful to human health. Limited studies suggest that the uptake of toxic heavy metals, notably Cd, by plants from PR is significantly lower than that from water-soluble P fertilizers produced from the same PR. Moreover, a PR source with a higher reactivity and Cd content can release more Cd than a PR with a lower reactivity and/or low Cd content for plant uptake. In addition to PR reactivity and Cd content, Cd uptake by plants also depends on soil properties, especially soil pH, and crop species.

In view of the fact that very limited information is available in the literature regarding secondary nutrients, micronutrients, liming effect, and hazardous elements associated with PR application, it is hoped that future research on PR will expand to include these areas so that the benefits and risks associated with PR use in soils and plants can be assessed more precisely.

The bulk of the world's PR production is utilized for P fertilizer manufacturing, thus only limited information on legislation for DAPR is available in literature. Regulations for the direct application of PRs should consider three main factors: PR reactivity (solubility), soil properties (mainly soil pH) and crop species. All current legislation on PR for direct application considers the quality of PR, namely: total P_2O_5 content, particle size distribution, and solubility. Issues associated with solubility measurements are complex and need careful consideration. It is difficult to develop universal legislation that all countries can adopt. Nevertheless, legislation to be adopted by a country or region should be based on and upgraded according to recent scientific findings in PR research. The guidelines proposed in this publication should provide useful information for establishing and revising legislation regarding PR for direct application.

Bibliography

Adams, M.A. & Pate, J.S. 1992. Availability of organic and inorganic forms of phosphorus to lupins (Lupinus spp.). *Plant Soil*, 145: 107–113.

Ae, N., Arihara, J., Okada, K., Yoshihara, T. & Johansen, C. 1990. Phosphorus uptake by pigeon pea and its role in cropping systems of the Indian subcontinent. *Science*, 248: 477–480.

Aigner, M., Fardeau, J.C. & Zapata, F. 2002. Does the P_i strip method allow assessment of the available soil P?: comparison against the reference isotope method. *Nutr. Cycl. Agroecosys.*, 63(1): 49–58.

Amberger, A. 1978. Experiences with soft rock phosphate for direct fertilizer application. *In* IFDC, ed. *Seminar on phosphate rock for direct application*, pp. 349–356. Muscle Shoals, Alabama, USA, IFDC.

Amer, F., Boulden, D.R., Black, C.A. & Duke, F.R. 1955. Characterization of soil phosphorus by anion exchange resin adsorption and P-32 equilibration. *Plant Soil*, 6: 391–408.

ANDA. 2001. *Statistics yearbook of the fertilizer sector*. Sao Paulo, Brazil, ANDA – National Association for Fertilizer Diffusion.

Anderson, D.L., Kussow, W.R. & Corey, R.B. 1985. Phosphate rock dissolution in soil: indications from plant growth studies. *Soil Sci. Soc. Am. J.*, 49: 918–925.

Anderson, G.C. & Sale, P.W.G. 1993. Application of the Kirk and Nye phosphate rock dissolution model. *Fert. Res.*, 35: 61–66.

Ankomah, A.B., Zapata, F., Danso, S.K.A. & Axmann, H. 1995. Cowpea varietal differences in uptake of phosphorus from Gafsa phosphate rock in a low P ultisol. Fert. Res., 41: 219 225.

Appleton, J.D. 2001. *Local phosphate resources for sustainable development in sub-Saharan Africa*. Keyworth, Nottingham, UK, National Environment Research Council, British Geological Survey and DFID.

Arihara, J. & Karasawa, T. 2000. Effect of previous crops on arbuscular mycorrhizal formation and growth of succeeding maize. *Soil Sci. Plant Nutr.*, 46: 43–51.

Attoe, O.J. & Olson, R.A. 1966. Factors affecting rate of oxidation in soils of elemental sulphur and that added in rock phosphate-sulphur fusions. *Soil Sci.*, 101: 317–324.

Axelrod, S. & Greidinger, D. 1979. Phosphate solubility test – interference of some accessory minerals. *J. Sci. Food Agric.*, 30:153–157.

Baanante, C.A. 1998. Economic evaluation of the use of phosphate fertilizers as a capital investment. *In* A.E. Johnston & J.K. Syers, eds. *Nutrient management for sustainable agriculture in Asia*, pp. 109–120. Wallingford, UK, CAB International.

Baanante, C.A. & Hellums, D.T. 1998. An analysis of the potential demand for phosphate fertilizers: sources of change and projections to 2025. *In* L.D. Currie & P. Loganathan, eds. *Long-term nutrient needs for New Zealand's primary industries: global supply, production requirements and environmental constraints*, pp. 7–31. Occasional Report No. 11. Palmerston North, New Zealand, Fertilizer and Lime Research Centre, Massey University.

Babare, A.M., Gilkes, R.J. & Sale, P.W.G. 1997. The effect of phosphate buffering capacity and other soil properties on North Carolina phosphate rock dissolution, availability of dissolved phosphorus and relative agronomic effectiveness. *Aus. J. Exp. Agri.*, 8: 845–1098.

Bachik, A.T. & Bidin, A., eds. 1992. *Proceedings of a workshop on phosphate sources for acid soils in the humid tropics of Asia.* 6–7 November 1990. Kuala Lumpur, Malaysian Society of Soil Science.

Bagayoko, M. & Coulibaly, B.S. 1995. Promotion and evaluation of Tilemsi phosphate rock in Mali agriculture. *In* H. Gerner & A.U. Mokwunye, eds. *Use of phosphate rock for sustainable agriculture in West Africa*, p.77–83. Miscellaneous Fertilizer Studies No. 11. Muscle Shoals, USA, IFDC Africa.

Bagyaraj, D.J. 1990. Ecology of VA mycorrhizae. *In* D.K. Arora, R. Bharat, K.G. Mukerji & G.R. Knudsen, eds. *Handbook of applied mycology, Vol. I. Soil and Plants*, pp. 3–34. New York, USA, Marcel Dekker Inc.

Bagyaraj, D.J. & Padmavathi Ravindra, T. 1993. Mycorriza. *In* P.K. Thampan, ed. *Organics in soil health and crop production.* Cochin, India, Peekay Tree Crops Developments Foundation.

Baligar, V.C., Fageria, N.K. & Ze, Z.L. 2001. Nutrient use efficiency in plants. *Com. Soil Sci. Plant Anal.*, 32: 921–950.

Bangar, K.C., Yadav, K.S. & Mishra, M.M. 1985. Transformation of rock phosphate during composting and the effect of humic acid. *Plant Soil*, 85: 259–266.

Barber, S.A. 1995. *Soil nutrient bioavailability. A mechanistic approach.* New York, USA, John Wiley & Sons.

Barea, J.M., Azcon, R. & Azcon-Aguilar, C. 1983. Interactions between phosphate solubilising bacteria and VA mycorrhiza to improve plant utilization of rock phosphate in non-acidic soils. *In* IMPHOS, ed. *3rd international congress on phosphorus compounds*, pp. 127–144. Brussels.

Barea, J.M., Toro, M., Orozco, M.E., Campos, E. & Azcon, R. 2002. The application of isotopic (^{32}P and ^{15}N) dilution techniques to evaluate the interactive effect of phosphate-solubilising rhizobacteria, mycorrhizal fungi and Rhizobium to improve the agronomic efficiency of rock phosphate for legume crops. *Nutr. Cycl. Agroecosys.*, 63 (1): 35–42.

Barnes, J.S. & Kamprath, E.J. 1975. Availability of North Carolina rock phosphate applied to soils. *N. C. Agric. Exp. St. Tech. Bull.*, 229. Raleigh, USA.

Basak, R.K., Karmakar, M. & Debnath, N.C. 1988. Fertilizer value of partially acidulated Purulia rock phosphate. *J. Ind. Soc. Soil Sci.*, 36: 729–732.

Bationo, A., Baethgen, W.E., Christianson, C.B. & Mokwunye, A.U. 1991. Comparison of five soil-testing methods to establish phosphorus sufficiency levels in soil fertilized with water-soluble and sparingly soluble phosphorus sources. *Fert. Res.*, 28: 271–279.

Baudet, G., Truong, B., Fayard, C. & Sustrac, G. 1986. La filière phosphate: du minéral à l'engrais, principaux points de repère. *Chron. Rech. Min.*, 484: 19–36.

Bekele, T., Cino, B.J., Ehlert, P.A.I., Van der Maas, A.A. & Van Diest, A. 1983. An evaluation of plant-borne factors promoting the solubilization of alkaline rock phosphates. *Plant Soil*, 75: 361–378.

Besoain, E., Rojas, C. & Montenegro, A., eds. 1999. *Las rocas fosforicas y su posibilidad de uso agricola en Chile.* Santiago, Instituto Nacional de Investigaciones Agricolas, Ministerio de Agricultura. 328 pp.

Bever, J.D., Schultz, P.A., Pringle, A. & Morton, J.B. 2001. Arbuscular mycorrhizal fungi: more diverse than meets the eye, and the ecological tale of why. *Bio. Sci.*, 51: 923–931.

Black, C.A. 1968. *Soil-plant relationships.* New York, USA, John Wiley & Sons.

Bojinova, D., Velkova, R., Grancharov, I. & Zhelev, S. 1997. The bioconversion of Tunisian phosphorite using Aspergillus niger. *Nut. Cyc. Agroecosys.*, 47: 227–232.

Bolan, N.S. & Hedley, M.J. 1990. Dissolution of phosphate rocks in soils. 2. Effect of pH on the dissolution and plant availability of phosphate rock in soil with pH dependent charge. Fert. Res., 24: 125 134.

Bolan, N.S. & Robson, D. 1987. Effects of vesicular-arbuscular mycorrhiza on the availability of iron phosphates to plants. *Plant Soil*, 99: 401–410.

Bolan, N.S., White, R.E. & Hedley, M.J. 1990. A review of the use of phosphate rocks as fertilizers for direct application in Australia and New Zealand. *Aus. J. Exp. Agric.*, 30: 297–313.

Bolan, N.S., Hedley, M.J., Syers, J.K. & Tillman, R.W. 1987 Single superphosphate-reactive phosphate rock mixtures. 1. Factors affecting chemical composition. *Fert. Res.*, 13: 223–239.

Bolland, M.D.A. & Gilkes, R.J. 1997. The agronomic effectiveness of reactive phosphate rocks, 2. Effect of phosphate rock reactivity. *Aus. J. Exp. Agric.*, 37: 937–946.

Bolland, M.D.A., Clarke, M.F. & Yeates, J.S. 1995. Effectiveness of rock phosphate, coastal superphosphate and single superphosphate for pasture on deep sandy soils. *Fert. Res.*, 41: 129–143.

Bolland, M.D.A., Lewis, D.C., Gilkes, R.J. & Hamilton, L.J. 1997. Review of Australian phosphate rock research. *Aus. J. Exp. Agric.*, 37: 845–859

BPPT-BRGM-CIRAD-TECHNIFERT-SPIE BATIGNOLLES. 1989. *The phosphates of the Ciamis and Tuban Regions (Java - Indonesia)*. Jakarta, Ministry of Research. 110 pp.

Braithwaite, A.C., Eaton, A.C. & Groom, P.S. 1989. Some factors associated with the use of the extractants 2% citric acid and 2% formic acid as estimators of available phosphorus in fertilizer products. *Fert. Res.*, 19: 175–181.

Braithwaite, A.C., Eaton, A.C. & Groom, P.S. 1990. Factors affecting the solubility of phosphate rock residues in 2% citric acid and 2% formic acid. *Fert. Res.*, 23: 37–42.

Bray, R.M. & Kurtz, L.T. 1945. Determination of total, organic and available forms of phosphorus in soils. *Soil Sci.*, 59: 39–45

British Sulphur Corporation Limited. 1987. *World survey of phosphate deposits*. London. 247 pp.

Brobst, D.A. & Pratt, W.P. 1973. *United States mineral resources*. U.S. Geological Survey Professional Paper 820.

Brundrett, M.C. 2002. Co-evolution of roots and mycorrhizas of land plants. *New Phyt.*, 154: 275–304.

Buresh, R.J., Smithson, P. & Hellums, D.T. 1997. Building soil phosphorus capital in Africa. *In* R.J. Buresh, P.A. Sanchez & F. Calhoun, eds. *Replenishing soil fertility in Africa*, pp. 111–149. Madison, USA, SSSA Special Publication 51.

Casanova, E. 1992. Las rocas fosfóricas naturales y modificadas y su uso potencial en suelos y cultivos de Venezuela. *In* Palmaven-Impofos, eds. *Memorias del curso sobre fertilización balanceada*. Valencia, Venezuela. 88 pp.

Casanova, E. 1993. Las rocas fosfóricas y su uso agroindustrial en Venezuela. Technical Notes. Palmaven, Venezuela. 124 pp.

Casanova, E. 1995. Agronomic evaluation of fertilizers with special reference to natural and modified phosphate rock. *Fert. Res.*, 41: 211–218.

Casanova, E. 1998. Suelos y fertilización de forrajes en Venezuela. *In* R. Tejos, C. Zambrano, L. Mancilla, W. García, M. Camargo, eds. *IV Seminario Manejo y utilización de pastos y forrajes en sistemas de producción animal*, pp. 129–136. Venezuela.

Casanova, E., ed. 1998. *Manejo eficiente de los fertilizantes fosfatados con énfasis en rocas fosoforicas de aplicación directa. Technical manual*. FAO/OIEA Regional Project of Technical Cooperation in Latin America (ARCAL). Maracay, Venezuela. 91 pp.

Casanova, E. & Lopez Perez A., eds. 1991. Alternativas como fertilizantes para los depósitos de fosfato de la América Tropical y Subtropical. *Rev. Fac. Agron. UCV*, Vol. 17, Nos. 1–4.

Casanova, E., Salas, A.M. & Toro, M. 1998. Manejo eficiente de los recursos fosfatados con énfasis en rocas fosfóricas de aplicación directa en Venezuela. *In* E. Casanova, ed. *Manejo eficiente de los fertilizantes fosfatados con énfasis en rocas fosfóricas de aplicación directa*, pp. 75–91. Proyecto ARCAL. Maracay, Venezuela.

Casanova, E., Salas, A.M. & Toro, M. 2002a. Evaluating the effectiveness of phosphate fertilizers in some Venezuelan soils. *Nutr. Cycl. Agroecosys.*, 63(1): 13–20.

Casanova, E., Salas, A.M. & Toro, M. 2002b. The use of nuclear and related techniques for evaluating the agronomic effectiveness of phosphate fertilizers, in particular rock phosphate in Venezuela. I. Phosphorus uptake, utilization and agronomic effectiveness. In *Assessment of soil P status and management of phosphatic fertilizers to optimize crop production*, pp. 93–100. Vienna, IAEA TECDOC 1272 IAEA. 473 pp.

Casanova, E., Goitia, R., Pereira, P., Comerma, J. & Aguilar, C. 1993. Necesidades y perspectivas agronómicas de fertilizantes y enmiendas en Venezuela. Venesuelos, 11: 17–23.

Castellano, M.A. & Molina, R. 1989. Mycorrhizae. *In* T.D. Landis, R.W. Tinue, S.E. MacDonald & J.P. Barnett, eds. Biological component: nursery pests and mycorrhizae, Vol. 5. *The container tree nursery manual*, pp 101–167. Agriculture Handbook 674. Washington, DC, USDA.

Castillo, J., Domínguez, J. & Barreiro, I. 1998. *Programa de cálculo de estimación de consumos de materias primas en la producción de RPA con roca Riecito y mezclas de ácidos fosfóricos y sulfúrico*. INTEVEP-PEQUIVEN Bulletin. 11p.

Chalk, P.M., Zapata, F. & Keerthisinghe, G. 2002. Towards integrated soil, water and nutrient management in cropping systems: the role of nuclear techniques. *In* IUSS, ed. Soil science: confronting new realities in the 21st century. Transactions 17th World Congress of Soil Science, CD ROM. p. 2164/1–2164–11. Bangkok.

Chardon, W.J., Menon, R.G. & Chien, S.H. 1996. Iron oxide-impregnated filter paper (Pi test): a review of its development and methodological research. *Nutr. Cycl. Agroecosys.*, 46: 41–51.

Chew, K.L. 1992. Distribution and acceptance of phosphate fertilizers: a Malaysian experience. *In* A.T. Bachik & A. Bidin, eds. *Proceedings of the workshop on phosphate sources for acid soils in the humid tropics of Asia*, p. 170–176. Kuala Lumpur, Malaysian Society of Soil Science.

Chien, S.H. 1977a. Dissolution of phosphate rocks in a flooded acid soil. *Soil Sci. Soc. Am. J.*, 41: 1106–1109.

Chien, S.H. 1977b. Thermodynamic considerations on the solubility of phosphate rock. *Soil Sci.*, 123: 117–121.

Chien, S.H. 1978. Interpretation of Bray I extractable phosphorus from acid soil treated with phosphate rock. *Soil Sci.*, 144: 34–39.

Chien, S.H. 1982. Direct application of phosphate rocks in some tropical soils of South America: a status report. *In* E. Pushparajah & S.H.A. Hamid, eds. *Phosphorus and potassium in the tropics*, pp. 519–529. Kuala Lumpur, Malaysian Society of Soil Science.

Chien, S.H. 1992. Reactions of phosphate rocks with acid soils of the humid tropics. *In* A.T. Bachik & A. Bidin, eds. *Proceedings of a workshop on phosphate sources for acid soils in the humid tropics of Asia*, pp. 18–29. Kuala Lumpur, Malaysian Society of Soil Science.

Chien, S.H. 1993. Solubility assessment for fertilizer containing phosphate rock. *Fert. Res.*, 35: 93–99.

Chien, S.H. 1995. Chemical, mineralogical, and solubility characteristics of phosphate rock for direct application. *In* K. Dahanayke, S.J. Van Kauwenbergh & D.T. Hellums, eds. *Direct application of phosphate rock and appropriate technology fertilizers in Asia: what hinders acceptance and growth*, pp. 49–58. Kandy, Sri Lanka, Institute of Fundamental Studies, and Muscle Shoals, USA, IFDC.

Chien, S.H. 1998. Evaluation of Gafsa (Tunisia) and Djebel Onk (Algeria) phosphate rocks and soil testing of phosphate rock for direct application. *In* A.E. Johnston & J.K. Syers, eds. *Nutrient management for sustainable food production in Asia*, pp. 175–185. Proc. IMPHOS-AARD/CSAR. Wallingford, UK, CAB International.

Chien, S.H. 2003. Factors affecting the agronomic effectiveness of phosphate rock: a general review. *In* S.S.S. Rajan & S.H. Chien, eds. *Direct application of phosphate rock and related technology: latest developments and practical experiences*. Proc. Int. Meeting, Kuala Lumpur, 16–20 July 2001. Muscle Shoals, USA, IFDC. 441 pp.

Chien, S.H. 2003. IFDC's evaluation of modified phosphate rock products. In *Proceedings of international meeting on direct application of phosphate rock and related technology: latest developments and practical experiences*. Kuala Lumpur, Malaysian Society of Soil Science, and Muscle Shoals, USA, IFDC.

Chien, S.H. & Black, C.A. 1976. Free energy of formation of carbonate apatites in some phosphate rocks. *Soil Sci. Soc. Am. J.*, 40: 234–239.

Chien, S.H. & Friesen, D.K. 1992. Phosphate rock for direct application *In* F.J. Sikora, ed. *Future directions for agricultural phosphorus research*, pp. 47–52. Bulletin Y-224. Muscle Shoals, USA, Valley Authority.

Chien, S.H. & Friesen, D.K. 2000. Phosphate fertilizers and management for sustainable crop production in tropical acid soils. *In* IAEA, ed. *Management and conservation of tropical acid soils for sustainable crop production*, pp. 73–89. IAEA-TECDOC-1159. Vienna, IAEA.

Chien, S.H. & Hammond, L.L. 1978. A comparison of various laboratory methods for predicting the agronomic potential of phosphate rock for direct application. *Soil Sci. Soc. Am. J.*, 42: 1758–1760.

Chien, S.H. & Hammond, L.L. 1988. *Agronomic evaluation of partially acidulated phosphate rocks in the tropics: IFDC experience*. IFDC-P-7. Muscle Shoals, USA, IFDC.

Chien, S.H. & Hammond, L.L. 1989. Agronomic effectiveness of partially acidulated phosphate rock as influenced by soil phosphorus-fixing capacity. *Plant Soil*, 120: 159–164.

Chien, S.H. & Menon, R.G. 1995a. Agronomic evaluation of modified phosphate rock products: IFDC's experience. *Fert. Res.*, 41: 197–209.

Chien, S.H. & Menon, R.G. 1995b. Factors affecting the agronomic effectiveness of phosphate rock for direct application. *Fert. Res.*, 41: 227–234.

Chien, S.H. & Van Kauwenbergh, S.J. 1992. Chemical and mineralogical characteristics of phosphate rock for direct application. *In* R.R. Campillo, ed. *First national seminar on phosphate rock in agriculture*, pp 3–31. Serie Carillanca No. 29. Temuco, Chile, Instituto de Investigaciones Agropecuarias.

Chien, S.H., Clayton, W.R. & McClellan, G.M. 1980a. Kinetic dissolution of phosphate rocks in soils. *Soil Sci. Soc. Am. J.*, 44: 260–264.

Chien, S.H., Hammond, L.L. & Leon, L.A. 1987b. Long term reactions of phosphate rocks with an oxisol in Colombia. *Soil Sci.*, 144: 257–265.

Chien, S.H., Leon, L.A. & Tejeda, H. 1980b. Dissolution of North Carolina phosphate rock in acid Colombian soils as related to soil properties. *Soil Sci. Soc. Am. J.*, 44: 1267–1271.

Chien, S.H., Menon, R.G. & Billingham, K.S. 1996 Phosphorus availability from phosphate rock as enhanced by water-soluble phosphorus. *Soil Sci. Soc. Am. J.*, 60: 1173–1177.

Chien, S.H., Sale, P.W.G. & Friesen, D.K. 1990a. A discussion of the methods for comparing the relative effectiveness of phosphate fertilizers varying in solubility. *Fert. Res.*, 24: 149–157.

Chien, S.H., Sale, P.W.G. & Hammond, L.L. 1990b. Comparison of effectiveness of various phosphate fertilizer products. In *Proceedings of international symposium on phosphorus requirements for sustainable agriculture in Asia and Oceania*, pp. 143–156. Manila, IRRI.

Chien, S.H., Adams, F., Khasawneh, F.E. & Henao, J. 1987a. Effects of combinations of triple superphosphate and a reactive phosphate rock on yield and phosphorus uptake by corn. *Soil Sci. Soc. Am. J.*, 51: 1656–1658.

Chien, S.H., Singh, U., Van Reuler, H. & Hellums, D.T. 1999. Phosphate rock decision support systems for sub-Saharan Africa. Special issue on Soil Fertility. *Afr. Fert. Mark.*, 12: 15–22.

Chien, S.H., Sompongse, D., Henao, J. & Hellums, D.T. 1987c. Greenhouse evaluation of phosphorus availability from compacted phosphate rocks with urea or with urea and triple superphosphate. *Fert. Res.*, 14: 245–256.

Comerma, J. & Paredes, R. 1978. Principales limitaciones y potencial agrícola de las tierras en Venezuela. *Agron. Trop.*, 28: 71–85.

Cooke, G.W. 1956. The value of phosphate rock for direct application. Emp. J. Exp. Ag., 24: 295 306.

Dabin, B. 1967. *Méthode Olsen modifiée*. Cahiers ORSTOM, Pédologie 5.3.

Dahanayake, K., Van Kauwenbergh, S.J. and Hellums, D.T., eds. 1995. *Direct application of phosphate rock and appropriate technology fertilizers in Asia: what hinders acceptance and growth.* Kandy, Sri Lanka, Institute of Fundamental Studies, and Muscle Shoals, USA, IFDC.

Date, R.A., Grundon, N.J., Rayment, G.E. & Probert, M.E., eds. 1995. *Plant-soil interactions at low pH: principles and management. developments in plant and soil sciences*. Vol. 64. Dordrecht, The Netherlands, Kluwer Academic Publishers. 822 pp.

De la Fuente-Martinez, J.M., Ramirez-Rodriguez, V., Cabrera-Ponce, J.L. & Herrera-Estrella, L. 1997. Aluminum tolerance in transgenic plants by alteration of citrate synthesis. *Science*, 276: 1566–1568.

Diamond, R.B. 1979. Views on marketing of phosphate rock for direct application. In IFDC, ed. Seminar on phosphate rock for direct application. Special Publication SP-1. Muscle Shoals, USA, IFDC.

Dominguez, J. & Barreiro, I. 1998. *Manual de instrucciones de trabajo para la operación de la planta RPA del complejo petroquímico Morón*. 63 pp.

Duddridge, J.A., Malibari, A. & Read, D.J. 1980. Structure and function of mycorrhizal rhizomorphs with special reference to their role in water transport. *Nature*, 287: 834–836.

Engelstad, O.P. & Hellums, D.T. 1993. *Water solubility of phosphate fertilizers: agronomic aspects – a literature review*. IFDC Paper Series P-17. Muscle Shoals, USA, IFDC.

Engelstad, O.P., Jugsujinda, A. & De Datta, S.K. 1974. Response by flooded rice to phosphate rocks varying in citrate solubility. *Soil Sci. Soc. Am. Proc.*, 38: 524–529.

Enyong, L.A., Debrah, S.K. & Bationo, A. 1999. Farmers' perceptions and attitudes towards introduced soil-fertility enhancing technologies in western Africa. *Nutr. Cycl. Agroecosys.*, 53(2): 177–187.

FAO. 1984. *Fertilizer and plant nutrition guide*. FAO Fertilizer and Plant Nutrition Bulletin No. 9. Rome.

FAO. 1995a. *Integrated plant nutrition systems*, by R. Dudal & R.N. Roy, eds. FAO Fertilizer and Plant Nutrition Bulletin No. 12. Rome.

FAO. 1995b. *World agriculture: towards 2010*, by N. Alexandratos, ed. New York, USA, John Wiley & Sons.

FAO. 1998. *Guide to efficient plant nutrient management*. Land and Water Development Division, Rome.

FAO. 1999. *Fertilizer yearbook 1998*. FAO Statistics Series No. 150. Rome.

FAO. 2001a. *Soil and nutrient management in sub-Saharan Africa in support of the Soil Fertility Initiative*. Land and Water Development Division. AGL/MISC/31/01. Rome.

FAO. 2001b. *Evaluation de l'efficacité agronomique des phosphates naturels pour l'application directe en Afrique de l'Ouest: synthèse des resultats agronomiques experimentaux*, by G. Bizimungu, R.N. Roy & W. Burgos. Rome. 49 pp.

FAO. *FAOSTAT statistics database* (available at http://apps.fao).

Fardeau, J.C. 1981. *Cinétique de dilution isotopique et phosphore assimilable des sols.* Univ. Paris VI. (Doctoral thesis)

Fardeau, J.C. 1993. Le phosphore assimilable des sols: sa représentation par un modèle à plusieurs compartiments. *Agronomie*, 13: 317–331.

Fardeau, J.C. 1996. Dynamics of phosphate in soils: an isotopic outlook. *Fert. Res.*, 45: 91–100.

Feder, G., Just, R.E. & Zilberman, D. 1985. Adoption of agricultural innovations in developing countries: a survey. *Econ. Dev. Cult. Ch.*, 33: 255–298.

Fixen, P.E. & Grove, J.H. 1990. Testing soils for phosphorus. *In* R.L. Westerman, ed. *Soil testing and plant analysis*, pp. 141–180. Book Series No. 3. Madison, USA, Soil Sci. Soc. Am. Inc.

Flach, E.N., Quak, W. & Van Diest, A. 1987. A comparison of the rock phosphate-mobilising capacities of various crop species. *Trop. Agric.*, 64: 347–352.

Formoso, M.L.L. 1999. *Workshop on tropical soils.* Rio de Janeiro, Brazil, Brazilian Academy of Sciences. 192 pp.

Fox, R.L. & Kamprath, E.J. 1970. Phosphorus sorption isotherms for evaluating the phosphate requirements of soils. *Soil Sci. Soc. Am. Proc.*, 34: 902–907.

Fox, R.L., Saunders, W.M.H. & Rajan, S.S.S. 1986. Phosphorus nutrition of pasture species: phosphorus requirement and root saturation values. *Soil Sci. Soc. Am. J.*, 50: 142–148.

Frederick, T., Truong, B. & Fayard, F. 1992. *Pre-feasibility study: production of modified phosphate fertilizers using Kodjari phosphate rock, Burkina Faso.* IFDC-CIRAD-TECHNIFERT. 91 pp.

Frossard, E., Brossard, M., Hedley, M.J. & Metherell, A. 1995. Reactions controlling the cycling of P in soils. *In* H. Tiessen, ed. *Phosphorus in the global environment*, p. 104–141. New York, USA, John Wiley & Sons.

Gahoonia, T.S., Claassen, N. & Jungk, A. 1992. Mobilization of phosphate in different soils by ryegrass supplied with ammonium or nitrate. *Plant Soil*, 140: 241–248.

Gaur, A.C. 1990. *Phosphorus solubilising microorganisms as biofertilisers.* New Delhi, Omega Scientific Publ. 176 pp.

Gerner, H. & Baanante, C.A. 1995. Economic aspects of phosphate rock application for sustainable agriculture in West Africa. *In* H. Gerner & A.U. Mokwunye, eds. *Use of phosphate rock for sustainable agriculture in West Africa*, pp. 134–141. Lomé, IFDC Africa.

Gerner, H. & Mokwunye, A.U., eds. 1995. *Use of phosphate rock for sustainable agriculture in West Africa*, pp. 77–83. Miscellaneous Fertilizer Studies No. 11. Muscle Shoals, USA, IFDC Africa.

Gillard, P, Sale, P.W.G. & Tennakoon, S.B. 1997. Building an expert system to advise on the use of reactive phosphate rock on Australian pasture. *Aus. J. Exp. Agric.*, 37: 1077–1084.

Goedert, W.J. 1983. Management of the Cerrado soils of Brazil; a review. *J. Soil Sci.* 34: 405–428.

Gremillion, L.R. & McClellan, G.H. 1975. Importance of chemical and mineralogical data in evaluating apatitic phosphate ores. Society of Mining Engineers of AIME. *Transactions*. Vol. 270.

Habib, L., Chien, S.H., Carmona, G. & Henao, J. 1999. Rape response to a Syrian phosphate rock and its mixture with triple superphosphate on a limed alkaline soil. *Com. Soil Sci. Plant Anal.*, 30: 449–456.

Habib, L., Chien, S.H., Menon, R.G. & Carmona, G. 1998. Modified iron oxide-impregnated paper strip test for soils treated with phosphate fertilizers. *Soil Sci. Soc. Am. J.*, 62: 972–976.

Halder, A.K., Mishra, A.K., Bhattacharyya, P. & Chakrabartty, P.K. 1990. Solubilization of rock phosphate by Rhizobium and Bradyrhizobium. *J. Gen. App. Microbiol.*, 36: 81–92.

Hammond, L.L. 1977. *Effectiveness of phosphate rocks in Colombian soils as measured by crop response and soil phosphorus levels*. Michigan State University, USA. (Ph.D. thesis)

Hammond, L.L. & Day, D.P. 1992. Phosphate rock standardization and product quality. *In* A.T. Bachik & A. Bidin, eds. *Proceedings of a workshop on phosphate sources for acid soils in the humid tropics of Asia*, pp. 73–89. Kuala Lumpur, Malaysian Society of Soil Science.

Hammond, L.L. & Leon, L.A. 1983. Agronomic effectiveness of natural and altered phosphate rocks from Latin America. *In* IMPHOS, ed. 3rd international congress on phosphorus compounds, pp. 503–518. Brussels.

Hammond, L.L., Chien, S.H. & Easterwood, G.W. 1986a. Agronomic effectiveness of Bayovar phosphate rock in soil with induced phosphorus retention. *Soil Sci. Soc. Am. J.*, 50: 1601–1606.

Hammond, L.L., Chien, S.H. & Mokwunye, A.U. 1986b. Agronomic value of unacidulated and partially acidulated phosphate rocks indigenous to the tropics. *Adv. Agron.*, 40: 89–140.

Hammond, L.L., Chien, S.H., Roy, A.H. & Mokwunye, A.U. 1989. Solubility and agronomic effectiveness of partially acidulated phosphate rocks as influenced by their iron and aluminium oxide content. *Fert. Res.*, 19: 93–98

Harjanto, S. 1986. *Phosphate deposits in Indonesia*. Workshop on occurrence, exploration and development of fertilizer minerals in the ESCAP region, 26–30 August 1986, Bangkok. pp. 191–197.

Haynes, R.J. 1992. Relative ability of a range of crop species to use phosphate rock and monocalcium phosphate as P sources when grown in soil. *J. Sci. Food Agric.*, 74: 1–7.

He, Z.L., Bian, W. & Zhu, J. 2002. Screening and identification of microorganisms capable of utilizing phosphate adsorbed by goethite. *Com. Soil Sci. Plant Anal.*, 33: 647–663.

Hedley, M.J. & Bolan, N.S. 1997. Developments in some aspects or reactive phosphate rock research and use in New Zealand. *Aus. J. Exp. Agric.*, 37: 861–884.

Hedley, M.J. & Bolan, N.S. 2003. Key outputs from reactive phosphate rock research in New Zealand. *In* S.S.S. Rajan & S.H. Chien, eds. *Direct application of phosphate rock and related technology: latest developments and practical experiences*. Proc. Int. Meeting, Kuala Lumpur, 16–20 July 2001. Muscle Shoals, USA, IFDC. 441 pp.

Hedley, M.J., Bolan, N.A. & Braithwaite, A.C. 1988. Single superphosphate-reactive phosphate rock mixtures. 2. The effect of phosphate rock type and denning time on the amounts of acidulated and extractable phosphate. *Fert. Res.*, 16: 179–194.

Hedley, M.J., Nye, P.H. & White, R.E. 1983. Plant induced changes in the rhizosphere of rape (Brassica napus cv Emerald) seedlings. IV. The effect of rhizosphere phosphorus status on the pH, phosphatase activity and depletion of soil phosphorus fractions in the rhizosphere and on the cation-anion balance in plants. *New Phyt.*, 95: 69–82.

Heerink, N., Kuyvenhoven, A. & Van Wijk, S.M. 2001. Economic policy reforms and sustainable land use in LDCs: issues and approaches. *In* N. Heerink, H. Van Keulen & M. Kuiper, eds. *Economic policy reforms and sustainable land use in LDCs. Recent advances in quantitative analysis*, pp. 1–20. Heidelberg, Germany, Physica Verlag.

Helleiner, G.K. 1975. Smallholder decision making: tropical African evidence. *In* L.G. Reynolds, ed. *Agriculture in development theory*. New Haven, USA, Yale University Press.

Helling, C.S., Chesters, G. & Corey, R.B. 1964. Contribution of organic matter and clay to soil cation-exchange capacity as affected by the pH of the saturating solution. *Soil Sci. Soc. Am. Proc.*, 28: 517–520.

Hellums, D.T. 1991. *Factors affecting the efficiency of nonconventional phosphorus fertilizers in lowland and upland cropping systems*. Auburn University, Auburn, USA. (Ph.D. thesis)

Hellums, D.T. 1992. Role of non-conventional phosphate fertilizers in tropical agriculture: IFDC's research perspective. *In* J.J. Schultz, ed. *Phosphate fertilizers and the environment*, pp. 89–95. Muscle Shoals, USA, IFDC.

Hellums, D.T. 1995a. Phosphorus and its role in crop production in Asia. *In* K. Dahanayake, S.J. Van Kauwenbergh & D.T. Hellums, eds. *Direct application of phosphate rock and appropriate technology fertilizers in Asia: what hinders acceptance and growth*, pp. 9–14. Kandy, Sri Lanka, Institute of Fundamental Studies, and Muscle Shoals, USA, IFDC.

Hellums, D.T. 1995b. Environmental aspects of phosphate fertilizer production and use. *In* K. Dahanayake, S.J. Van Kauwenbergh & D.T. Hellums, eds. *Direct application of phosphate rock and appropriate technology fertilizers in Asia: what hinders acceptance and growth*, pp. 105–114. Kandy, Sri Lanka, Institute of Fundamental Studies, and Muscle Shoals, USA, IFDC.

Hellums, D.T., Baanante, C.A. & Chien, S.H. 1992. Alternative phosphorus fertilizers for the tropics: an agronomic and economic evaluation. *In* S. Balas, ed. *Proceedings of the tropsoils phosphorus decision support system workshop*, pp. 147–154. Raleigh, USA, Tropical Soils 92-01.

Hellums, D.T., Chien, S.H. & Touchton, J.T. 1989. Potential agronomic value of calcium in some phosphate rocks from South America and West Africa. *Soil Sci. Soc. Am. Proc.*, 53: 459–462.

Helyar, K.R. 1998. Efficiency of nutrient utilization and sustaining soil fertility with particular reference to phosphorus. *Field Crops Res.*, 56: 187–195.

Henao, J. & Baanante, C. A. 1999. *An evaluation of strategies to use indigenous and imported sources of phosphorus to improve soil fertility and land productivity in Mali*. Technical Bulletin IFDC-T-49. Muscle Shoals, USA, IFDC. 75 pp.

Heng, L.K. 2000. Modelling, database and the P submodel. *In* IAEA, ed. *Management and conservation of tropical acid soils for sustainable crop production*, pp. 101–111. IAEA-TECDOC-1159. Vienna, IAEA.

Heng, L.K. 2003. Towards developing a decision support system for phosphate rock direct application in agriculture. In *Proceedings of the international meeting on direct application of phosphate rock and related technology: latest developments and practical experiences*. Special Publication. Muscle Shoals, USA, IFDC.

Hien, V., Kaboré, D., Youl, S. & Löwenberg de Boer, J. 1997. Stochastic dominance analysis of on-farm trial data: the riskiness of alternative phosphate sources in Burkina Faso. *Ag. Econ.*, 15(3): 213–221.

Hinsinger, P. 1998. How do plant roots acquire mineral nutrients? *Adv. Agron.*, 64: 225–265.

Hinsinger, P. & Gilkes, R.J. 1995. Root-induced dissolution of phosphate rock in the rhizosphere of lupins grown in alkaline soil. *Aus. J. Soil. Res.*, 33: 477–489.

Hinsinger, P. & Gilkes, R.J. 1997. Dissolution of phosphate rock in the rhizosphere of five plant species grown in an acid, P-fixing substrate. *Geoderma*, 75: 231–249.

Hocking, P.J. 2001. Organic acids exuded from roots in phosphorus uptake and aluminum tolerance of plants in acid soils. *Adv. Agron.*, 74: 63–93.

Hocking, P.J., Keerthisinghe, G., Smith, F.W. & Randall, P.J. 1997. Comparison of the ability of different crop species to access poorly-available soil phosphorus. *In* T. Ando, K. Fujita, K.H. Matsumoto, S. Mori & J. Sekiya, eds. *Plant nutrition for sustainable food production and agriculture*, pp. 305–308. Dordrecht, The Netherlands, Kluwer Academic Publishers.

Hocking, P.J., Randall, P., Delhaize, E. & Keerthisinghe, G. 2000. The role of organic acids exuded from roots in phosphorus nutrition and aluminum tolerance in acidic soils. *In* IAEA, ed. *Management and conservation of tropical acid soils for sustainable crop production*, pp. 61–73. IAEA-TECDOC-1159. IAEA, Vienna.

Hoffland, E. 1992. Quantitative evaluation of the role of organic acid exudation in the mobilization of rock phosphate by rape. *Plant Soil*, 140: 279–289.

Hoffland, E., Findenegg, G.R. & Nelemans, J.A. 1989. Solubilization of rock phosphate by rape. II. Local root exudation of organic acids as a response to P-starvation. *Plant Soil*, 113: 161–165.

Horst, W.J & Waschkies, C. 1987. Phosphorus nutrition of spring wheat in mixed culture with white lupin. *Z. Plaenz. Bodenk.*, 150: 1–8.

Howeler, R.H. & Woodruff, C.M. 1968. Dissolution and availability to plants of rock phosphates of igneous and sedimentary origins. *Soil Sci. Soc. Am. Proc.*, 32: 79–82.

Hu, H.Q., Li, X.Y., Liu, J.F., Liu, F.L., Xu, L. & Liu, F. 1997. The effect of direct application of phosphate rock on increasing crop yield and improving properties of red soil. *Nutr. Cycl. Agroecosys.*, 46: 235–239.

Hulse, J.H. 1995. *Science, agriculture and food security*. Ottawa, NRC Research Press.

IAEA. 2000. *Management and conservation of tropical acid soils for sustainable crop production*. Proceedings of a consultants meeting, 1–3 March 1999. IAEA-TECDOC-1159. Vienna.

IAEA. 2002. *Assessment of soil phosphorus status and management of phosphatic fertilizers to optimize crop production*. IAEA-TECDOC-1272. Vienna. (also available on CD ROM)

IFDC. 1996. *Pilot plant test to produce granular partially acidulated phosphate rock products by the single step acidulation granulated process using Riecito phosphate rock*. Final Report. Muscle Shoals, USA.

IFDC. 2003. *Direct application of phosphate rock and related technology: latest experiences and practical experiences,* S.S.S. Rajan & S.H. Chien, eds. Proc. Int. Meeting, Kuala Lumpur, 16–20 July 2001. Muscle Shoals, USA, International Fertilizer Development Center. 441 pp.

IFDC-CIRAD-ICRAF-NORAGRIC. 1996. An assessment of phosphate rock as a capital investment: evidence from Burkina Faso, Madagascar, Zimbabwe. Washington, DC, World Bank. 23 pp.

Illmer, P. & Scinner, F. 1995. Solubilization of inorganic calcium phosphates – solubilisation mechanisms. *Soil Biol. Biochem.*, 27: 257–263.

Illmer, P., Barbato, A. & Scinner, F. 1995. Solubilization of hardly-soluble $AlPO_4$ with P solubilising microorganisms. *Soil Biol. Biochem.*, 27: 265–270.

IMPHOS. 1983. *Third international congress on phosphorus compounds*. Proc. Congress, 4–6 October 1983, Brussels. 656 pp.

IMPHOS. 1992. *Fourth international IMPHOS conference*. Proc. Int. Conf., Ghent, Belgium. 758 pp.

Iretskaya, S.N. & Chien, S.H. 1999. Comparison of cadmium uptake by five different food grain crops grown on three soils varying in pH. *Com. Soil Sci. Plant Anal.*, 30: 441–448.

Iretskaya, S.N., Chien, S.H. & Menon, R.G. 1998. Effect of acidulation of high cadmium containing phosphate rocks on cadmium uptake by upland rice. *Plant Soil*, 201:183–188.

Ishikawa, S., Wagatsuma, T., Sasaki, R. & Ofei-Manu, P. 2000. Comparison of the amount of citric and malic acids in Al media of seven plant species and two cultivars each in five plant species. *Soil Sci. Plant Nutr.*, 46: 751–758.

Jadin, P. & Truong, B. 1987. Efficacité de deux phosphates naturels tricalciques dans deux sols ferrallitiques acides du Gabon. Café Cacao Thé, Vol. XXXI, no. 4: 291–302.

Jaggi, T.N. 1986. Potentiality of using ground rock phosphates as phosphatic fertilizers in Indian soils. *In* G.V. Kothandaraman, T.S. Manickam & K. Natarajan, eds. *Rock phosphate in agriculture*, pp 1–14. Coimbatore, India, Tamil Nadu Agricultural University.

Jama, B., Swinkels, A. & Buresh, R.J. 1997. Agronomic and economic evaluations of organic and inorganic phosphorus in western Kenya. *Agron. J.*, 89: 597–604.

Johnston, A.E & Syers, J.K., eds. 1996. *Nutrient management for sustainable food production in Asia.* Proc. Int. Conf. IMPHOS-AARD/CSAR. Wallingford, UK, CAB International.

Johnston, H.W. 1954a. The solubilization of "insoluble" phosphate. II – A quantitative and comparative study of the action of selected aliphatic acids on tricalcium phosphate. *N. Z. J. Sci. Tech. B.*, 36: 49–55.

Johnston, H.W. 1954b. The solubilization of "insoluble" phosphate. III – A quantitative and comparative study of the action of chosen aromatic acids on tricalcium phosphate. *N. Z. J. Sci. Tech. B.*, 36: 281–284.

Johnstone, P.D. & Sinclair, A.G. 1991. Replication requirements in field experiments for comparing phosphate fertilizers. *Fert. Res.*, 29: 329–333.

Jones, D.L. 1998. Organic acids in the rhizosphere – a critical review. *Plant Soil*, 205: 25–44.

Jones, M.D., Duraqll, D.M. & Tinker, P.B. 1998. A comparison of arbuscular and ectomycorrhizal Eucalyptus coccifera: growth response, phosphorus uptake efficiency, and external hyphal production. *New Phyt.*, 140: 125–134.

Kamh, M., Horst, W.J., Amer, F., Mostafa, H. & Maier, P. 1999. Mobilization of soil and fertilizer phosphate by cover crops. *Plant Soil*, 211: 19–27.

Kamprath, E.J. 1970. Exchangeable aluminum as a criterion for liming leached mineral soils. *Soil. Sci. Soc. Am. Proc.*, 34: 252–254.

Kanabo, I. & Gilkes, R.J. 1987. The role of soil pH in the dissolution of phosphate rock fertilizers. *Fert. Res.*, 12: 165–174.

Keerthisinghe, G., Zapata, F., Chalk, P.M. & Hocking, P. 2001. Integrated approach for improved P nutrition of plants in tropical acid soils. *In* W.J. Horst, M.K. Schenk, A. Bürkert, N. Classen, H. Flessa, W.B. Frommer, H. Goldbach, H.W. Olfs, V. Römheld, B. Sattelmacher, U. Schmidhalter, S. Schubert, N. v. Wiren & L. Wittenmayer, eds. *Food security and sustainability of agro-ecosystems through basic and applied research*, pp. 974–975. Proceedings of the XIV International Plant Nutrition Colloquium. Dordrecht, The Netherlands, Kluwer Academic Publishers.

Khasawneh, F.E. & Doll, E.C. 1978. The use of phosphate rock for direct application to soils. *Adv. Agron.*, 30: 159–206

Kirk, G.J.D. & Nye, P.H. 1985a. The dissolution and dispersion of dicalcium phosphate dihydrate in soils. I. A predictive model for a planar source. *J. Soil Sci.*, 36: 445–459.

Kirk, G.J.D. & Nye, P.H. 1985b. The dissolution and dispersion of dicalcium phosphate dihydrate in soils. II. Experimental evaluation of the model. *J. Soil Sci.*, 36: 461–468.

Kirk, G.J.D. & Nye, P.H. 1986. The dissolution and dispersion of dicalcium phosphate dihydrate in soils. III. A predictive model for regularly distributed particles. *J. Soil Sci.*, 37: 511–524.

Kittams, H.A. & Attoe, O.J. 1965. Availability of phosphorus in rock phosphate-sulphur fusions. *Agron. J.*, 57: 331–334

Klockner Industrie Anlagen GMBH. 1968. Etude economique et technique en vue de l'exploitation d'un gisement de phosphates dans la région de Bourem, Mali. Bamako, Dir. Nat. Geol. Mines. 63 pp.

Kpomblekou, K. & Tabatabai, M.A. 1994a. Effect of organic acids on release of phosphorus from phosphate rocks. *J Soil Sci.*, 158: 442–453.

Kpomblekou, K. & Tabatabai, M.A. 1994b. Metal content of phosphate rocks. *Com. Soil Sci. Soc. Plant Anal.*, 25: 2871–2882.

Kpomblekou, K, Chien, S.H., Henao, J. & Hill, W.A. 1991. Greenhouse evaluation of phosphate fertilizers produced from Togo phosphate rock. *Com. Soil Sci. Plant Anal.*, 22: 63–73.

Kucey, R.M.N., Janzen, H.H. & Leggett, M.E. 1989. Microbially mediated increases in plant-available phosphorus. *Adv. Agron.*, 42: 199–228.

Kuyvenhoven, A., Becht, J.A. & Ruben, R. 1998a. Critères economiques d'un investissement public dans l'amélioration du sol en Afrique de l'Ouest: prévisions d'utilisations du phosphate naturel pour l'amélioration de la fertilité du sol. *In* H. Breman & K. Sissoko, eds. *Intensification agricole du Sahel*, pp. 895–916. Paris, Karthala.

Kuyvenhoven, A., Becht, J.A. & Ruben, R. 1998b. Financial and economic evaluation of phosphate rock use to enhance soil fertility in West Africa: is there a role for government? *In* G.A.A. Wossink, G.C. Van Kooten & G.H. Peters, eds. *Economics of agro-chemicals*, pp. 249–261. Aldershot, UK, Ashgate.

Kuyvenhoven, A., Ruben, R. & Kruseman, G. 1998c. Technology, market policies and institutional reform for sustainable land use in southern Mali. *Ag. Econ.*, 19: 53–62.

Lal, R. 1990. Tropical soils: distribution, properties and management. *Res. Man. Opt.*, 7: 39–52.

Lal, R. 1999. Soil management and restoration for C sequestration to mitigate the accelerated greenhouse effect. *Prog. Env. Sci.*, 1: 307–326.

Lal, R. 2000. Soil management in the developing countries. *Soil Sci.*, 165(1): 57–72.

Lange Ness, R. & Vlek, P.L.G. 2000. Mechanism of calcium and phosphate release from hydroxy-apatite by mycorrhizal hyphae. *Soil Sci. Soc. Am. J.*, 64: 949–955.

Larsen, S. 1952. The use of ^{32}P in studies of the uptake of phosphorus by plants. *Plant Soil*, 4: 1–10.

Ledgard, S.F., Thorrold, B.S., Sinclair, A.G., Rajan, S.S.S. & Edmeades, E.C. 1992. Summary of MAF field trials with 'Longlife' phosphatic fertilizers. *Proc. N. Z. Grass. Ass.*, 54: 35–40.

Lehr, J.R. & McClellan, G.H. 1972. *A revised laboratory reactivity scale for evaluating phosphate rocks for direct application.* Bulletin Y-43. Muscle Shoals, USA, Tennessee Valley Authority.

Lenglen, M. 1935. *La question de l'emploi direct des phosphates minéraux naturels.* Paris, Syndicat National des Engrais Chimiques. 46 p.

Leon, L.A., Fenster, W.E. & Hammond, L.L. 1986. Agronomic potential of eleven phosphate rocks from Brazil, Colombia, Peru, and Venezuela. *Soil Sc. Soc. Am. J.*, 50(3): 798–802.

Ling, A.H., Harding, P.E. & Ranganathan, V. 1990. Phosphorus requirements and management of tea, coffee and cocoa. *In* IRRI, ed. *Phosphorus requirements for sustainable agriculture in Asia and Oceania*, pp. 383–398. Manila, IRRI.

Loganathan, P., Hedley, M.J. & Bretherton, M.R. 1994. The agronomic value of co-granulated Christmas Island grade C phosphate rock and elemental sulphur. *Fert. Res.*, 39: 229–237.

Lompo, F., Sedogo, M.P. & Hien, V. 1995. Agronomic impact of Burkina phosphate and dolomite limestone. *In* H. Gerner & A.U. Mokwunye, eds. *Use of phosphate rock for sustainable agriculture in West Africa*, pp. 54–66. Miscellaneous Fertilizer Studies No. 11. Muscle Shoals, USA, IFDC Africa.

Lopes, A. 1998. The use of phosphate rocks to build up soil P and increase food production in acid soils: the Brazilian experience. *In* A.E. Johnston & J.K. Syers, eds. *Nutrient management for sustainable food production in Asia*, pp. 121–132. Proc. IMPHOS-AARD/CSAR. Wallingford, UK, CAB International.

Mackay, A.D. & Syers, J.K. 1986. Effect of phosphate, calcium, and pH on the dissolution of a phosphate rock in soil. *Fert. Res.*, 10: 175–184.

Mackay, A.D., Syers, J.K & Gregg, P.E.H. 1984. Ability of chemical extraction procedures to assess the agronomic effectiveness of phosphate rock materials. *N. Z. J. Agric. Res.*, 27: 219–230.

Maene, L.M. 2003. Direct application of phosphate rock: a global perspective of the past, present, and future. *In* S.S.S. Rajan & S.H. Chien, eds. *Direct application of phosphate rock and related technology: latest developments and practical experiences.* Proc. Int. Meeting, Kuala Lumpur, 16–20 July 2001. Muscle Shoals, USA, IFDC. 441 pp.

Mahimairaja, S., Bolan, N.S. & Hedley, M.J. 1995. Dissolution of phosphate rock during the composting of poultry manure: an incubation experiment. *Fert. Res.*, 40: 93–104.

Malaysian Standard. 1998. *Specification for ground phosphate rock, agricultural grade (second revision), MS46: 1998.* Selangor DE, Malaysia, Department of Standards.

Manickam, T.S. 1993. Organics in soil fertility and productivity management. *In* P.K. Thampan, ed. *Organics in soil health and crop production*, pp. 87–104. Cochin, India, Peekay Tree Crops Developments Foundation.

Manjunath, A. & Habte, M. 1992. External and internal P requirements of plant species differing in their mycorrhizal dependency. *A. Soil Res. Rehab.* 6: 271–284.

Marschner, H. 1993. *Mineral nutrition of higher plants*. London, Academic Press Ltd., Harcourt Brace & Co. Publishers.

Matthews, R., Stephens, W., Hess, T., Middleton, T. & Graves, A. 2002. Applications of crop/soil simulation models in tropical agricultural systems. *Adv. Agron.*, 76: 31–124.

McClellan, G.H. 1980. Mineralogy of carbonate fluorapatites. J. Geol. Soc., 6: 675 681.

McClellan, G.H. & Gremillion, L.R. 1980. Evaluation of phosphatic raw materials. *In* F.E. Khasawneh, ed. *The role of phosphorus in agriculture*, pp. 43–80. Madison, USA, ASA-SSSA.

McClellan, G.H. & Lehr, J.R. 1969. Crystal chemical investigation of natural apatites. Am. Min., 54: 1374 1391.

McClellan, G.H. & Notholt, A.F.G. 1986. Phosphate deposits of tropical sub-Saharan Africa. *In* A.E. Mokwunye & P.L.G. Vlek, eds. *Management of nitrogen and phosphorus fertilizers in sub-Saharan Africa*, pp. 173–224. Developments in Plant and Soil Sciences 24. Dordrecht, The Netherlands, Kluwer Academic Publishers.

McClellan, G.H. & Van Kauwenbergh, S.J. 1990a. Mineralogy of sedimentary apatites. *In* A.J.G. Notholt & I. Jarvis, eds. *Phosphorite research and development*. Geological Society Special Publication 52: 23–31.

McClellan, G.H. & Van Kauwenbergh, S.J. 1990b. Relationship of mineralogy to sedimentary phosphate rock reactivity. *In* A.T. Bachik & A. Bidin, eds. *Proceedings of the workshop on phosphate sources for acid soils in the humid tropics of Asia*, pp. 1–17. Kuala Lumpur, Malaysian Society of Soil Science.

McClellan, G.H. & Van Kauwenbergh, S.J. 1991. Mineralogical and chemical variation of francolites with geological time. *J. Geol. Soc.*, 148: 809–812.

McConnell, D. 1938. A structural investigation of the isomorphism of the apatite group. *Am. Min.*, 54: 1379–1391.

McLaughlin, M.J., Simpson, P., Fleming, F., Stevens, D.P., Cozens, G. & Smart, M.K. 1997. Effect of fertilizer type on cadmium and fluorine concentrations in clover herbage. *Aus. J. Exp. Ag.*, 37: 1019–1026.

McLay, C.D., Rajan, S.S.S. & Liu, Q. 2000. Agronomic effectiveness of partially acidulated phosphate rock fertilizers in an allophonic soil at near-neutral pH. *Com. Soil Sci. Plant Anal.*, 31: 423–435.

Menge, J.A., Lembright, H. & Johnson, E.L.V. 1977. Utilization of mycorrhizal fungi in citrus nurseries. *Proc. Int. Soc. Cit.*, 1: 129-132.

Menon, R.G. & Chien, S.H. 1990. Phosphorus availability to maize from partially acidulated phosphate rocks and phosphate rocks compacted with triple superphosphate. *Plant Soil*, 127: 123–128.

Menon, R.G. & Chien, S.H. 1995. Soil testing for available phosphorus in soils where phosphate rock-based fertilizers are used. *Fert. Res.*, 41: 179–181.

Menon, R.G. & Chien, S.H. 1996. *Compaction of phosphate rocks with soluble phosphates – an alternative technology to partial acidulation of phosphate rocks with low reactivity: IFDC's experience.* IFDC-T-44. Muscle Shoals, USA, IFDC.

Menon, R.G., Chien, S.H. & Chardon, W.J. 1997. Iron oxide-impregnated filter paper (Pi test): II. A review of its application. *Nut. Cyc. Agroecosys.*, 47: 7–18.

Menon, R.G., Chien, S.H. & Hammond, L.L. 1989a. The P_i soil phosphorus test: a new approach to testing for soil phosphorus. IFDC Reference Manual R-7. Muscle Shoals, USA, IFDC.

Menon, R.G., Chien, S.H. & Hammond, L.L. 1990. Development and evaluation of the P_i soil test for plant-available phosphorus. *Com. Soil Sci. Plant Anal.*, 21: 1131–1150.

Menon, R.G., Hammond, L.L. & Sissingh, H.A. 1989b. Determination of plant-available phosphorus by the iron hydroxide-impregnated paper (P_i) soil test. *Soil Sci. Soc. Am. J.*, 53: 110–115.

Metherell, A.K. & Perrott, K.W. 2003. An integrated decision support package for evaluation of reactive phosphate rock fertilizer strategies for grazed pasture. *In* S.S.S. Rajan & S.H. Chien, eds. *Direct application of phosphate rock and related technology: latest developments and practical experiences.* Proc. Int. Meeting, Kuala Lumpur, 16–20 July 2001. Muscle Shoals, USA, IFDC. 441 pp.

Mew, M. 2000. Phosphate rock. In *Metals and mineral annual review*, pp. 110–122. London, The Mining Journal Ltd.

Mishra, B. 1975. Report on investigations on the utilization of Mussoorie rock phosphate for direct application to field crops. *In* PPCL, ed. *Mussoorie-phos, a phosphate fertilizer for direct application.* Dehradun, India, PPCL. 34 pp.

Mishra, M.M. & Bangar, K.C. 1986. Rock phosphate composting: transformation of phosphorus forms and mechanisms of solubilization. *Bio. Ag. Hort.*, 3: 331–340.

Miyasaka, C. & Habte, M. 2001. Plant mechanisms and mycorrhizal symbioses to increase phosphorus uptake efficiency. *Com. Soil Sci. Plant Anal.*, 32: 1101–1147.

Moghimi, A. & Tate, M.E. 1978. Does 2-Ketogluconate chelate calcium in the pH range 2.4–6.4? *Soil Biol. Biochem.*, 10: 289–292.

Mokwunye, A.E. & Vlek, P.L.G., eds. 1986. *Management of nitrogen and phosphorus fertilizers in sub-Saharan Africa*, pp. 173–224. Developments in Plant and Soil Sciences 24. Dordrecht, The Netherlands, Kluwer Academic Publishers.

Montange, D. & Zapata, F. 2002. Standard characterization of soils employed in the FAO/IAEA phosphate project. *In* IAEA, ed. *Assessment of soil phosphorus status and management of phosphatic fertilizers to optimise crop production*, pp. 24–43. IAEA TECDOC 1272. Vienna, IAEA. 473 pp.

Montenegro, A. & Zapata, F. 2002. Rape genotypic differences in P uptake and utilization from phosphate rocks in an andisol of Chile. *Nut. Cyc. Agroecosys.*, 63(1): 27–33.

Moorby, H., White, R. & Nye, P. 1988. The influence of phosphate nutrition on H ion efflux from the roots of young rape plants. *Plant Soil*, 105: 247–256.

Morel, C. & Fardeau, J.C. 1989. Native soil and fresh fertilizer phosphorus uptake as affected by rate of application and P fertilizers. *Plant Soil*, 115: 123–128.

Mortvedt, J.J. & Sikora, F.J. 1992. Heavy metal, radionuclides, and fluorides in phosphorus fertilizers. *In* F.J. Sikora, ed. *Future directions for agricultural phosphorus research*, pp. 69–73. TVA Bulletin Y-224. Muscle Shoals, USA.

Mosse, B., Stribley, D.P. & Le Tacon, F. 1981. Ecology of mycorrhizae and mycorrhizal fungi. *Adv. Micro. Ecol.*, 5: 137–210.

Mostara, M.R. & Datta, N.P. 1971. Rock phosphate as a fertilizer for direct application in acid soils. *J. Ind. Soil Sci.*, 19: 107–113.

Mugwira, L., Nyamangara, J. & Hikwa, D. 2002. Effect of manure and fertilizer on maize at a research station and in a smallholder (peasant) area of Zimbabwe. *Plant Soil*, 33: 379–402.

Murdoch, C.L., Jacobs, J.A. & Gerdemann, J.W. 1967. Utilization of phosphorus sources of different availability by mycorrhizal and non-mycorrhizal maize. *Plant Soil*, 27: 329–338.

Nakamaru, Y., Nanzyo, M. & Yamasaki, S. 2000. Utilization of apatite in fresh volcanic ash by pigeonpea and chickpea. *Soil Sci. Plant Nut.*, 46: 591–600.

Nelson, W.L., Mehlich, A. & Winters, E. 1953. The development, evaluation, and use of soil tests for phosphorus availability. *Agron. J.*, 4: 153–158.

Netherlands Economic Institute (NEI). 1998. *Opérationnalisation d'un programme d'utilisation du Burkina phosphate*. Restricted circulation. Rotterdam, The Netherlands.

Nordengren, S. 1957. New theories of phosphate reactions in the soil. *Fert. Feed. Stuffs J.*, 47: 344–353

Notholt, A.J.G., Sheldon, R.P. & Davidson, D.F., eds. 1989. *Phosphate deposits of the world, Vol. 2: phosphate rock resources*. Cambridge, UK, University Press Cambridge.

Official Journal of the European Communities. 1976. No. L24, 30.1.76, p.29.

Oldeman, L.R. 1994. The global extent of soil degradation. *In* D.J. Greenland & I. Szabolcs, eds. *Soil resilience and sustainable land use*, pp. 99–118. Wallingford, UK, CAB International.

Olsen, S.R., Cole, C.V., Watanabe, F.S. & Dean, L.A. 1953. *Estimation of available phosphorus in soils by extraction with sodium bicarbonate*. USDA Circ. 939.

Ortas, I., Harris, P.J. & Rowell, D.L. 1996. Enhanced uptake of phosphorus by mycorrhizal sorghum plants as influenced by forms of nitrogen. *Plant Soil*, 184: 255–264.

Ortas, I., Ortakci, D. & Kaya, Z. 2002. Various mycorrhizal fungi propagated on different hosts have different effect on citrus growth and nutrient uptake. *Com. Soil Sci. Plant Anal.*, 33: 259–272.

Owusu-Bennoah, E., Zapata, F. & Fardeau, J.C. 2002. Comparison of greenhouse and P isotopic laboratory methods for evaluating the agronomic effectiveness of natural and modified rock phosphates in some acid soils of Ghana. *Nut. Cyc. Agroecosys.*, 63(1): 1–12.

Pairunan, A.K., Robson, A.D. & Abbott, L.K. 1980. The effectiveness of vesicular-arbuscular mycorrhizae in increasing growth and phosphorus uptake of subterranean clover from phosphorus sources of different solubilities. *New Phyt.*, 84: 327–338.

Palaniappan, S.P. & Natarajan, K. 1993. Practical aspects of organic matter maintenance in soil. *In* P.K. Thampan, ed. *Organics in soil health and crop production*, pp. 24–41. Cochin, India, Peekay Tree Crops Development Foundation.

Pearson, R.W. 1975. *Soil acidity and liming in the humid tropics*. Cornell International Agriculture Bulletin 30. Ithaca, New York, Cornell University. 66 pp.

Perrott, K. 2003. Direct application of phosphate rocks to pastoral soils – phosphate rock reactivity and the influence of soil and climatic factors. *In* S.S.S. Rajan & S.H. Chien, eds. *Direct application of phosphate rock and related technology: latest developments and practical experiences*. Proc. Int. Meeting, Kuala Lumpur, 16–20 July 2001. Muscle Shoals, USA, IFDC. 441 pp.

Perrott, K.W. & Wise, R.G. 2000. Determination of residual reactive phosphate rock in soil. *Com. Soil Sci. Plant Anal.*, 31: 1809–1824.

Perrott, K.W., Saggar, S. & Menon, R.G. 1993. Evaluation of soil phosphate status where phosphate rock based fertilizers have been used. *Fert. Res.*, 35: 67–82.

Perrott, K.W., Kerr, B.E., Watkinson, J.H. & Waller, J.E. 1996. Phosphorus status of pastoral soils where reactive phosphate rock fertilizers have been used. *Proc. N. Z. Grass Ass.*, 57: 133–137.

Pessarakli, M., ed. 1999. Handbook of plant and crop stress. New York, USA, Marcel Dekker Inc. 1254 pp.

Pohlman, A.A. & McColl, G.J. 1986. Kinetics of metal dissolution from forest soils by organic acids. *J. Env. Qual.*, 15: 86–92.

Poojari, B.T., Krishnappa, K.M., Sharma, K.M.S., Jayakumar, B.V. & Panchaksharaiah, S. 1988. Efficiency of rock phosphate as a source of phosphorus in rice-groundnut cropping system in coastal Karnataka. In *Seminar proceedings on the use of rock phosphate in west coast soils*, pp. 58–63. Bangalore, India, University of Agricultural Sciences.

Pushparajah, E., Cnah, F. & Magat, S.S. 1990. Phosphorus requirements and management of oil palm, coconut and rubber. *In* IRRI, ed. *Phosphorus requirements for sustainable agriculture in Asia and Oceania*, pp. 399–425. Manila, IRRI.

Pyrites, Phosphates and Chemicals Ltd. (PPCL). 1980. *Mussoorie phosphorite as straight phosphatic fertilizer*. Dehradun, India.

Pyrites, Phosphates and Chemicals Ltd. (PPCL). 1982. *Mussoorie-phos, a phosphate fertilizer for direct application*. Mussoorie Phosphorite Project. Dehradun, India.

Pyrites, Phosphates and Chemicals Ltd. (PPCL). 1983. *Research on Mussoorie phosphate rock*. Technical Bulletin No. 1. New Delhi.

Quin, B.F. & Scott, P. 2003. Development of the market for direct application phosphate rock a perspective based on experience in New Zealand and Scotland. *In* S.S.S. Rajan & S.H. Chien, eds. *Direct application of phosphate rock and related technology: latest developments and practical experiences*. Proc. Int. Meeting, Kuala Lumpur, 16–20 July 2001. Muscle Shoals, USA, IFDC. 441 pp.

Rajan S.S.S. 1973. Phosphorus adsorption characteristics of Hawaiian soils and their relationships to equilibrium concentrations required for maximum growth of millet. *Plant Soil*, 39: 519–532

Rajan, S.S.S. 1982a Availability to plants of phosphate from "biosupers" and partially acidulated phosphate rock. *N. Z. J. Ag. Res.*, 25: 355–361.

Rajan, S.S.S. 1982b. Influence of phosphate rock reactivity and granule size on the effectiveness of 'biosuper'. *Fert. Res.*, 3: 3–12.

Rajan, S.S.S. 1983. Effect of sulphur content of phosphate rock/sulphur granules on the availability of phosphate to plants. *Fert. Res.*, 4: 287–296.

Rajan, S.S.S. 1987. Phosphate rock and phosphate rock/sulphur granules as phosphate fertilizers and their dissolution in soil. *Fert. Res.*, 11: 43–60.

Rajan, S.S.S. 2002. Comparison of phosphate fertilizers for pasture and their effect on soil solution phosphate. *Com. Soil Sci. Plant Anal.*, 33: 2227–2245.

Rajan, S.S.S. & Ghani, A. 1997. Differential influence of soil pH on the availability of partially sulphuric and phosphoric acidulated phosphate rocks. II. Chemical and scanning electron microscopic studies. *Nut. Cyc. Agroecosys.*, 48: 171–178.

Rajan, S.S.S. & Marwaha, B.C. 1993. Use of partially acidulated phosphate rocks as phosphate fertilisers. *Fert. Res.*, 35: 47–59.

Rajan, S.S.S. & Watkinson, J.H. 1992. Unacidulated and partially acidulated phosphate rock: agronomic effectiveness and the rates of dissolution of phosphate rock. *Fert. Res.*, 33: 267–277.

Rajan, S.S.S., Watkinson, J.H. & Sinclair, A.G. 1996. Phosphate rock for direct application to soils. *Ad. Agron.*, 57: 78–159.

Rajan, S.S.S., Brown, M.W., Boyes, M.K. & Upsdell, M.P. 1992. Extractable phosphorus to predict agronomic effectiveness of ground and unground phosphate rocks. *Fert. Res.*, 32: 291–302.

Rajan, S.S.S., Fox, R.L., Saunders, W.M.H. & Upsdell, M.P. 1991a. Influence of pH, time and rate of application on phosphate rock dissolution and availability to pastures. I. Agronomic benefits. *Fert. Res.,* 28: 85–93.

Rajan, S.S.S., Fox, R.L., Saunders, W.M.H. & Upsdell, M.P. 1991b. Influence of pH, time and rate of application on phosphate rock dissolution and availability to pastures. II. Soil chemical studies. *Fert. Res.*, 28: 95–101.

Rao, I.M., Friesen, D.K. & Osaki, M. 1999. Plant adaptation to phosphorus-limited tropical soils. *In* M. Pessarakli, ed. *Handbook of plant and crop stress*, pp. 61–96. New York, USA, Marcel Dekker Inc.

Ratkowsky, D.A., Tennakoon, S.B., Sale, P.W.G. & Simpson, P.G. 1997. The use of substitution values for characterizing fertilizer performance. *Aus. J. Exp. Ag.* 37: 913–920.

Ratsimbazafy, J.R. 1975. *Reconnaisssance préliminaire des dépôts phosphatés des Iles Barren*. Doc. A.2223. Tananarive, Madagascar, Service Geologique de Madagascar.

Rauniyar, G.P. & Goode, F.M. 1992. Technology adoption on small farms. *World Dev.*, 20(2): 275–282.

RELARF. 1996. *Resumenes de comunicaciones orales*. IV Reunion de la Red Latinoamericana de Roca Fosfórica, 3–5 julio 1996. Havana.

Robinson, J.S. & Syers, J.K. 1991. Effects of solution calcium concentration and calcium sink size on the dissolution of Gafsa phosphate rock in soils. *J. Soil Sci.*, 42: 389–397.

Robinson, J.S., Syers, J.K. & Bolan, N.S. 1992. Importance of proton supply and calcium-sink size in the dissolution of phosphate rock materials of different reactivity in soil. *J. Soil Sci.*, 43: 447–459.

Rodriguez, R. & Herrera, J. 2002. Field evaluation of partially acidulated phosphate rocks in a ferralsol from Cuba. *Nut. Cyc. Agroecosys.*, 63(1): 43–48.

Rogers, E.M. 1983. *Diffusion of innovations*. New York, USA, The Free Press.

Rogers, H.T., Pearson, R.W. & Ensminger, L.E. 1953. Soil and fertilizer phosphorus in plant nutrition. New York, USA, Academic Press.

Rong, M. 1995. Phosphate rocks and fertilizers in China. *In* K. Dahanayake, S.J. Van Kauwenbergh & D.T. Hellums, eds. *Direct application of phosphate rock and appropriate technology fertilizers in Asia – what hinders acceptance and growth*, pp. 187–189. Kandy, Sri Lanka, Institute of Fundamental Studies, and Muscle Shoals, USA, IFDC.

Saggar, S., Hedley, M.J. & White, R.E. 1990. A simplified resin membrane technique for extracting phosphorus from soils. *Fert. Res.*, 24: 173–180.

Saggar, S., Hedley, M.J. & White, R.E. 1992a. Development and evaluation of an improved soil test for phosphorus: 1. The influence of phosphorus fertilizer solubility and soil properties on the extractability of soil P. *Fert. Res.*, 33: 81–91.

Saggar, S., Hedley, M.J., White, R.E., Gregg, P.E., Perrott, K.W. & Comforth, I.S. 1992b. Development and evaluation of an improved soil test for phosphorus: 2. Comparing of the Olsen and mixed cation-anion exchange resin tests for predicting the yield of ryegrass grown in pots. *Fert. Res.*, 33: 135–144.

Sale, P.W.G. & Mokwunye, A.U. 1993. Use of phosphate rocks in the tropics. *Fert. Res.*, 35: 33–45.

Sale, P.W.G., Simpson, P.G., Anderson, C.A. & Muir, L.L., eds. 1997a. The role of reactive phosphate rocks fertilizers for pastures in Australia. *Aus. J. Exp. Ag.*, 37: 845–1023. Reprint: Melbourne, Australia, CSIRO Publishing.

Sale, P.W.G., Simpson, P.G., Lewis, D.C., Gilkes, R.J., Bolland, M.D.A., Ratkowsky, D.A., Gilbert, M.A., Garden, D.L., Cayley, J.W.D. & Johnson, D. 1997b. The agronomic effectiveness of reactive phosphate rocks: 1. Effect of the pasture environment. *Aus. J. Exp. Ag.*, 8: 921–936.

Sanchez, P.A. & Buol, S.W. 1975. Soils of the tropics and the world food crisis. *Science*, 188: 598–603.

Sanchez, P.A. & Salinas, J.G. 1981. Low-input technology for managing oxisols and ultisols in tropical America. *Adv. Agron.*, 34: 280–398.

Sari, N., Ortas, I. & Yetisir, H. 2002. Effect of mycorrhiza inoculation on plant growth, yield, and phosphorus uptake in garlic under field conditions. *Com. Soil Sci. Plant Anal.*, 33: 2189–2201.

Schnitzer, M. & Skinner, S.I.M. 1969. Free radicals in soil humic compounds. *Soil. Sci.*, 108: 383–388.

Schultz, J. 1986. Sulphuric acid-based partially acidulated phosphate rock: its production cost and use. IFDC-T-31. Muscle Shoals, USA, IFDC.

Secilia, J. & Bagyaraj, D.J. 1992. Selection of efficient vesicular-arbuscular mycorrhizal fungi for wetland rice (Oryza sativa L). *Bio. Fert. Soils*, 13: 108–111.

Sery, A. & Greaves, G.N. 1996. Chemical state of Cd in apatite phosphate ore as determined by EXAFS spectroscopy. *Am. Min.*, 81: 864–873.

Shapiro, B.I. & Sanders, J.H. 1998. Fertilizer use in semi-arid West Africa: profitability and supporting policy. *Ag. Sys.*, 56(4): 467–482.

Sheldon, R.P. 1987. Industrial minerals, with emphasis on phosphate rock. *In* D.J. McLaren & B.J. Skinner, eds. *Resources and world development*, pp. 347–361. New York, USA, John Wiley & Son Limited.

Siddique, M., Ghonsikar, C.P. & Malewar, G.U. 1986. Studies on the use of Mussoorie rock phosphate in combination with some indigenous solubilising materials on calcareous soil. *In* G.V. Kothandaraman, T.S. Manickam & K. Natarajan, eds. *Rock phosphate in agriculture*, pp. 142–149. Coimbatore, India, Tamil Nadu Agricultural University.

Sikora, F.J. 2002. Evaluating and quantifying the liming potential of phosphate rocks. *Nut. Cyc. Agroecosys.*, 63(1): 59–67.

Silverman, S.R., Fuyat, R.K. & Weiser, J.D. 1952. *Quantitative determination of calcite associated with carbonate-bearing apatites*, pp. 211–222. US Geological Survey.

Simpson, P.G. 1997. *Reactive phosphate rocks: their potential role as P fertilizer for Australian pastures.* Technical Bulletin. Melbourne, Australia, La Trobe University.

Sinclair, A.G., Shannon, P.W. & Risk, W.H. 1990. Sechura phosphate rock supplies plant-available molybdenum for pastures. *N. Z. J. Ag. Res.*, 33: 499–502.

Sinclair, A.G, Johnstone, P.D., Smith, L.C., O'Connor, M.B. & Nguyen, L. 1993a. Comparison of six phosphate rocks and single superphosphate as phosphate fertilizers for clover-based pasture. *N. Z. J. Ag. Sci.*, 41: 415–420.

Sinclair, A.G., Johnstone, P.D., Smith, L.C., O'Connor, M.B. & Nguyen, L. 1993b. Agronomy, modelling and economics of reactive phosphate rocks as slow-release phosphate fertilizers for grasslands. *Fert. Res.*, 36: 229–238.

Sinclair, A.G., Johnstone, P.D., Smith, L.C., Risk, W.H., O'Connor, M.B., Roberts, A.H., Morton, J.D., Nguyen, L. & Shannon, P.W. 1993c. Effect of reactive phosphate rock on the pH of soil under pasture. *N. Z. J. Agric. Res.*, 36: 381–384.

Singaram, P., Rajan, S.S.S. & Kothandaraman, G.V. 1995. Phosphate rock and a phosphate rock/superphosphate mixture as fertilizers for crops grown on a calcareous soil. *Com. Soil Sci. Plant Anal.*, 26: 1571–1583.

Singh, C.P. & Amberger, A. 1990. Humic substances in straw compost with rock phosphate. *Bio. Wastes*, 31: 165–174.

Singh, C.P. & Amberger, A. 1991. Solubilization and availability of phosphorus during decomposition of rock phosphate enriched straw and urine. *Bio. Ag. Hort.*, 7: 261–269.

Singh, S. & Kapoor, K.K. 1999. Inoculation with phosphate-solubilising microorganisms and a vesicular-arbuscular mycorrhizal fungus improves dry matter yield and nutrient uptake by wheat grown in a sandy soil. *Bio. Fert. Soils*, 28: 139–144.

Singh, U., Wilkens, P.W., Henao, J., Chien, S.H., Hellums, D.T. & Hammond, L.L. 2003. An expert system for estimating agronomic effectiveness of freshly applied phosphate rock. In *Direct application of phosphate rock and related appropriate technology – latest developments and practical experiences*. Special publication. Muscle Shoals, USA, IFDC.

Sissoko, K. 1998. *Et demain l'agriculture? Options techniques et mesures politiques pour un développement agricole durable en Afrique subsaharienne*. Documents sur la gestion des ressources tropicales. Wageningen, The Netherlands, Wageningen University.

Smith, F.W. & Grava, J. 1958. Availability of phosphorus contained in phosphatic shale compared to that contained in monocalcium phosphate and raw rock phosphate. *Soil Sci.*, 86: 313–318.

Smith, F.W., Ellis, B.G. & Grava, J. 1957. Use of acid-fluoride solutions for the extraction of available phosphorus in calcareous soils and in soils to which rock phosphate has been added. *Soil Sci. Soc. Am. Proc.*, 21: 400–404.

Sprague, R.H. Jr & Carlson, E.H. 1982. *Building effective decision support systems*. Englewood Cliffs, USA, Prentice-Hall Inc.

Sri Adiningsih, J. & Nassir, A. 2001. *The potential for improving crop production on upland acid soils in Asia through the use of appropriate phosphate rocks*. Proceedings of the World Fertilizer Congress: Fertilization in the Third Millenium: Fertilizer, Food Security and Environmental Protection, 3–9 August 2001, Beijing.

Srinivasan, T.N. 1994. Foreign trade policies and India's development. In *Agriculture and trade in China and India*, p. 177. San Francisco, USA, International Centre for Economic Growth, ICS Press.

Stevenson, F.J. 1967. Organic acids in soil. *In* D.A. McLaren & G.H. Peterson, eds. *Soil biochemistry*, pp. 119–146. New York, USA, Marcel Dekker Inc.

Stowasser, W.F. 1991. Phosphate rock – analysis of the phosphate rock situation in the United States: 1990–2040. *Eng. Min. J.*, 192(9): 16CC-16II.

Subba Rao, N.S. 1982a. Biofertilizers. *In* N.S. Subba Rao, ed. *Advances in agricultural microbiology*, pp. 219–242. Oxford and IBH, UK, Mohan Prilani, and New Delhi, Butterworth and Co.

Subba Rao, N.S. 1982b Utilization of farm wastes and residues in agriculture. *In* N.S. Subba Rao, ed. *Advances in agricultural microbiology*, pp. 509–522. Oxford and IBH, UK, Mohan Prilani, and New Delhi, Butterworth and Co.

Swaby, R.J. 1975. Biosuper – biological superphosphate. *In* K.D. McLachlan, ed. *Sulphur in Australasian agriculture*, pp. 213–220. Sydney, Australia, Sydney University Press.

Sylvia, D.M. 1992. Demonstration and mechanism of improved phosphorus uptake by vesicular-arbuscular mycorrhizal fungi. *In* F.J. Sikora, ed. *Future directions for agricultural phosphorus research*, pp. 31–34. Muscle Shoals, USA, National Fertilizer and Environmental Research Centre, TVA.

Tandon, H.L.S. 1987. *Phosphorus: research and agricultural production in India*. Fertilizer Development and Consultation Organization. New Delhi, Greater Kailash 1.

Tandon, H.L.S. 1991. *Sulphur research and agricultural production in India*. Washington, DC, The Sulphur Institute. 88 pp.

Teboh, J.F. 1995. Phosphate rock as a soil amendment: who should bear the cost? *In* H. Gerner & A.U. Mokwunye, eds. *Use of phosphate rock for sustainable agriculture in West Africa*, pp. 142–149. Lomé, IFDC Africa.

Toro, M., Azcon, R. & Barea, J.M. 1997. Improvement of arbuscular mycorrhiza development by inoculation of soil with phosphate-solubilising rhizobacteria to improve rock phosphate bioavailability (^{32}P) and nutrient cycling. *App. Env. Microbiol.*, 63: 4408–4412.

Truong, B. 1986. *Synthèse des résultats des essais phosphates du Togo, bruts et partiellement attaqués, de 1983 à 1985*. Rapport de mission d'appui à la Direction de la Recherche Agronomique, Lomé. 22 pp.

Truong, B. & Cisse, L. 1985. Appréciation de la valeur fertilisante des phosphates de Matam (Sénégal). *Agron. Trop.*, 40(3): 230–238.

Truong, B. & Fayard, C. 1987. *Proposition d'une filière d'engrais au Burkina Faso à base de phosphates naturels de Kodjari, partiellement solubilisés*. Etude de faisabilité pour le Ministère de l'Agriculture et de l'Elevage (Burkina Faso) et du Ministère de la Coopération (France). 90 pp.

Truong, B. & Fayard, C. 1988. *Proposition d'une approche raisonnée pour la valorisation des phosphates naturels du Vénézuela*. Rapport CIRAD-TECHNIFERT. 83 pp.

Truong, B. & Fayard, C. 1993. *Etude de prefaisabilité pour une production d'engrais au Mali à partir des phosphates de Tilemsi*. Rapport CIRAD-TECHNIFERT. Montpellier, France. 127 pp.

Truong, B. & Fayard, C. 1995. Small-scale fertilizer production units using raw and partially solubilized phosphate. *In* H. Gerner & A.U. Mokwunye, eds. *Use of phosphate rock for sustainable agriculture in West Africa*, pp. 181–198. Miscellaneous Fertilizer Studies No. 11. Muscle Shoals, Alabama, USA, IFDC Africa.

Truong, B. & Montange, D. 1998. The African experience with phosphate rock, including Djebel Onk , and case studies in Brazil and Vietnam. *In* A.E. Johnston & J.K. Syers, eds. *Nutrient management for sustainable food production in Asia*, pp. 133–148. Proc. IMPHOS-AARD/CSAR. Wallingford, UK, CAB International.

Truong, B. & Pichot, J. 1976. Influence du phosphore des graines de la plante test sur la détermination du phosphore isotopiquement diluable (valeur L). *Agron. Trop.*, 31(11): 379–386.

Truong, B. & Zapata, F. 2002. Standard characterization of phosphate rock samples from the FAO/IAEA phosphate project. *In* IAEA, ed. *Assessment of soil phosphorus status and management of phosphatic fertilizers to optimise crop production*, pp. 9–23. IAEA TECDOC. 1272. Vienna, IAEA. 473 pp.

Truong, B., Pichot, J. & Beunard, P. 1978. Caracterisation et comparaison des phosphates naturels tricalciques d'Afrique de l'Ouest en vue de leur utilisation directe en agriculture. *Agron. Trop.*, 33: 136–145.

Truong, B., Beunard, P., Diekola, K. & Pichot, J. 1982. Caractérisation et comparaison des phosphates naturels de Madagascar en vue de leur utilisation en agriculture. *Agron. Trop.*, 37(1): 9–16.

U.S. Bureau of Mines & U.S. Geological Survey. 1981. *Principles of a resource/reserve classification for minerals*. U.S. Geological Survey Circular 831.

U.S. Bureau of Mines. 2001. *Mineral commodity summaries (1981–2001), phosphate rock*. Washington, DC, U.S. Department of the Interior.

U.S. Geological Survey. 1982. *Sedimentary phosphate resource classification system of the U.S. Bureau of Mines and the U.S. Geological Survey*. U.S. Geological Survey Circular 882.

United Nations Environment Programme (UNEP). 2000. *Global environment outlook 2000*. London, Earthscan Publications Ltd.

United Nations Industrial Development Organization (UNIDO) & IFDC. 1998. *Fertilizer manual*. Dordrecht, The Netherlands, Kluwer Academic Publishers. 615 pp.

Valencia, I., Pieri, C. & Hellums, D.T. 1994. Rock phosphate as a capital investment in natural resource management. *In* ISSS and MSSS, eds. *Trans. 15th World Congress Soil Science*, Commission VI Symposia, Vol. 7a: 227–233. Acapulco, Mexico.

Van der Paauw, F. 1971. Effective water extraction method for the determination of plant available soil phosphorus. *Plant Soil*, 34: 467–481.

Van der Zee, S.E.A.T., Fokking, L.G.J. & Van Riemdijk, W.H. 1987. A new technique for assessment of reversibly adsorbed phosphate. *Soil Sci. Soc. Am. J.*, 51: 599–604.

Van Kauwenbergh, S.J. 1995. Mineralogy and characterization of phosphate rock. *In* K. Dahanayake, S.J. Van Kauwenbergh & D.T. Hellums, eds. *Direct application of phosphate rock and appropriate technology fertilizers in Asia – what hinders acceptance and growth*, pp. 29–47. Kandy, Sri Lanka, Institute of Fundamental Studies.

Van Kauwenbergh, S.J. 1997. *Cadmium and other minor elements in world resources of phosphate rock*. Proceedings No. 400. London, The Fertilizer Society.

Van Kauwenbergh, S.J. 2003. Overview of world phosphate rock production. *In* S.S.S. Rajan & S.H. Chien, eds. *Direct application of phosphate rock and related technology: latest developments and practical experiences*. Proc. Int. Meeting, Kuala Lumpur, 16–20 July 2001. Muscle Shoals, USA, IFDC. 441 pp.

Van Kauwenbergh, S.J. & Hellums, D.T. 1995. Direct application phosphate rock: a contemporary snapshot. *Phos. Pot.*, 200: 27–37.

Van Kauwenbergh, S.J. & McClellan, G.H. 1990a. Mineralogy of sedimentary apatites and the relationship to phosphate rock reactivity. In Proc. National Workshop on Fertilizer Efficiency, Cisarua, Indonesia. 29 pp.

Van Kauwenbergh, S.J. & McClellan, G.H. 1990b. Comparative geology and mineralogy of the southeastern United States and Togo phosphorites. *In* A.J.G. Notholt & I. Jarvis, eds. *Phosphorite research and development*, pp. 139–155. Geological Society Special Publication No. 52.

Vanlauwe, B., Nwoke, O.C., Diels, J., Sanginga, N., Carsky, R.J., Deckers, J. & Merckx, R. 2000. Utilization of rock phosphate by crops on a representative sequence in Northern Guinea savanna zone of Nigeria: response by Mucuna pruriens, Lablab purpureus and maize. *Soil Bio. Bioch.*, 32: 2063–2077.

Verma, L.N. 1993. Biofertiliser in agriculture. *In* P.K. Thampan, ed. *Organics in soil health and crop production*, pp. 152–183. Cochin, India, Peekay Tree Crops Development Foundation.

Von Uexkull, H.R. & Mutert, E. 1995. Global extent, development and economic impact of acid soils. *In* R.A. Date, N.J. Grundon, G.E. Rayment & M.E. Probert, eds. *Plant-soil interactions at low pH: principles and management*. Development in Plant and Soil Sciences 64. Dordrecht, The Netherlands, Kluwer Academic Publishers. p 5–19.

Wani, S.P. & Lee, K.K. 1992. Role of biofertilisers in upland crop production. *In* H.L.S. Tandon, ed. *Fertilisers, organic manures, recyclable wastes and biofertilisers*, pp. 91–112. New Delhi, Fertiliser Development and Consultation Organization.

Watkinson, J.H. 1994a. Modelling the dissolution of reactive phosphate rock in New Zealand pastoral soils. *Aus. J. Soil Res.*, 32: 739–53.

Watkinson, J.H. 1994b. A test for phosphate rock reactivity in which solubility and size are combined in a dissolution rate function. *Fert. Res.*, 39: 205–215.

Weil, R.R. 2000. Soil and plant influence on crop response to two African phosphate rocks. *Agron. J.*, 92: 1167–1175.

Weil, S., Gregg, P.E.H. & Bolan, N.S. 1994. Influence of soil moisture on the dissolution of reactive phosphate rocks. *In* L.D. Currie & P. Loganathan, eds. *The efficient use of fertilizers in a changing environment: reconciling productivity and sustainability*, pp. 75–81. Occasional Report No. 7. Palmerston North, New Zealand, Fertilizer and Lime Research Centre, Massey University.

World Bank. 1994. *Feasibility of phosphate rock as a capital investment in sub-Saharan Africa: issues and opportunities*. Washington, DC.

World Bank. 1997. *PR Initiative case studies: synthesis report. An assessment of phosphate rock as a capital investment: evidence from Burkina Faso, Madagascar, and Zimbabwe.* IFDC/CIRAD/ICARF/NORAGRIC. Washington, DC.

Yost, R.S., Naderman, G.C., Kamprath, E.J. & Lobata, E. 1982. Availability of rock phosphate as measured by an acid-tolerant pasture grass and extractable phosphorus. *Agron. J.*, 74: 462–46.

Young, C. 1990. Effect of phosphorus-solubilizing bacteria and vesicular-arbuscular mycorrhizal fungi on the growth of tree species in sub-tropical-tropical soils. *Soil Sci. Plant Nut.*, 36: 225–231.

Yusdar, H. & Hanafi, M. 2003. Use of phosphate rock for perennial and annual crops cultivation in Malaysia: a review. *In* S.S.S. Rajan & S.H. Chien, eds. *Direct application of phosphate rock and related technology: latest developments and practical experiences.* Proc. Int. Meeting, Kuala Lumpur, 16–20 July 2001. Muscle Shoals, USA, IFDC. 441 pp.

Zapata, F. 2000. Evaluating the agronomic effectiveness of phosphate rocks using nuclear and related techniques: results from a past FAO/IAEA Coordinated Research Project. *In* IAEA, ed. *Management and conservation of tropical acid soils for sustainable crop production*, pp. 91–100. IAEA-TECDOC-1159. Vienna, IAEA.

Zapata, F. 2003. FAO/IAEA research activities on direct application of phosphate rocks for sustainable crop production. *In* S.S.S. Rajan & S.H. Chien, eds. *Direct application of phosphate rock and related technology: latest developments and practical experiences.* Proc. Int. Meeting, Kuala Lumpur, 16–20 July 2001. Muscle Shoals, USA, IFDC. 441 pp.

Zapata, F., ed. 1995. Evaluation of the agronomic effectiveness of phosphate fertilizers through the use of nuclear and related techniques. Special issue. *Fert. Res.*, 41: 167–242.

Zapata, F., ed. 2002. Utilisation of phosphate rocks to improve soil status for sustainable crop production in acid soils. Special issue. *Nut. Cyc. Agroecosys.*, 63(1):1–98.

Zapata, F. & Axmann, H. 1995. ^{32}P isotopic techniques for evaluating the agronomic effectiveness of rock phosphate materials. *Fert. Res.*, 41: 189–195.

Zapata, F. & Zaharah, A.R. 2002. Phosphorus availability from phosphate rock and sewage sludge as influenced by the addition of water-soluble phosphate fertilizer. *Nut. Cyc. Agroecosys.*, 63(1): 43–48.

Zapata, F., Axmann, H. & Braun, H. 1986. Agronomic evaluation of rock phosphate materials by means of radioisotope techniques. *In* ISSS & BG, eds. *Trans. 13th Int. Cong. Soil Sci.*, Vol. III: 1012–1013. Hamburg, Germany.

Zapata, F., Casanova, E., Salas, A.M. & Pino, I. 1994. Dynamics of phosphorus in soils and phosphate fertilizer management in different cropping systems through the use of isotopic techniques. *In* ISSS & MSSS, eds. *Trans. 15th World Congress Soil Sci., Commission IV Symposia*, Vol. 5a: 451–466. Acapulco, Mexico.

Zapata, F., Pino, I., Baherle, P. & Parada, A.M. 1996. Estudio comparativo de la eficiencia de uso y absorcion de fosforo a partir de fertilizantes fosforicos por genotipos de trigo. *Terra*, 14: 325–330.

FAO TECHNICAL PAPERS

FAO FERTILIZER AND PLANT NUTRITION BULLETINS

1. Fertilizer distribution and credit schemes for small-scale farmers, 1979 (E* F)
2. Crop production levels and fertilizer use, 1981 (E* F S)
3. Maximizing the efficiency of fertilizer use by grain crops, 1980 (E F S)
4. Fertilizer procurement, 1981 (E F)
5. Fertilizer use under multiple cropping systems, 1983 (C* E F)
6. Maximizing fertilizer use efficiency, 1983 (E*)
7. Micronutrients, 1983 (C* E* F S*)
8. Manual on fertilizer distribution, 1985 (E* F)
9. Fertilizer and plant nutrition guide, 1984 (Ar C* E* F* S*)
10. Efficient fertilizer use in acid upland soils of the humid tropics, 1986 (E F S)
11. Efficient fertilizer use in summer rainfed areas, 1988 (E F S*)
12. Integrated plant nutrition systems, 1995 (E F)
13. Use of phosphate rocks for sustainable agriculture, 2003 (E F**)

Availability: November 2003

Ar	– Arabic	Multil +	Multilingual
C	– Chinese	*	Out of print
E	– English	**	In preparation
F	– French		
P	– Portuguese		
S	– Spanish		

The FAO Technical Papers are available through the authorized FAO Sales Agents or directly from Sales and Marketing Group, FAO, Viale delle Terme di Caracalla, 00100 Rome, Italy.